BONE ROOMS

BONE ROOMS

FROM SCIENTIFIC RACISM
to HUMAN PREHISTORY
in MUSEUMS

SAMUEL J. REDMAN

Harvard University Press

Cambridge, Massachusetts | London, England

2016

Copyright © 2016 by the President and Fellows of Harvard College
ALL RIGHTS RESERVED

Second printing

Library of Congress Cataloging-in-Publication Data
Redman, Samuel J., author.
  Bone rooms : from scientificr acism to human prehistory in museums / Samuel J. Redman.
    pages cm
  Includes bibliographical references and index. ISBN 978-0-674-66041-0
  1. Human remains (Archaeology)—United States. 2. Archaeological museums and collections—United States—History—19th century. 3. Archaeological museums and collections—United States—History—20th century.
  4. Archaeology—United States—History. 5. Racism in anthropology—United States—History. I. Title.
  CC79.5.H85R43 2016
  930.1074'73—dc23
                                                        2015033855

CONTENTS

Prologue · 1

1 · Collecting Bodies for Science · 16

2 · Salvaging Race and Remains · 69

3 · The Medical Body on Display · 126

4 · The Story of Man through the Ages · 158

5 · Scientific Racism and Museum Remains · 188

6 · Skeletons and Human Prehistory · 227

Epilogue · 277

Notes   293
Acknowledgments   355
Index   359

BONE ROOMS

PROLOGUE

Gunshots ripped through the late-spring air near a dusty U.S. Army outpost in rural Minnesota in May 1864. Militiamen who were engaged in a campaign against local Indians shot a Dakota man twice: one bullet struck him in the head, shattering his skull; the other tore through his mouth or neck.[1] Either wound alone could have been fatal. The man likely died instantly or bled to death in seconds. Healthy and strong in life, he now lay on the ground completely disfigured. Described in contemporary newspaper accounts as a "hostile Sioux"—and later by scientists as a man of distant Asiatic descent—he was probably between twenty-five and thirty-five years old. A single incident such as this, even a deadly one, on the distant Minnesota frontier might have soon vanished from memory in a nation focused on violent clashes with Native Americans across the region and the raging Civil War miles away. What happened to the body of this particular young Dakota man, however, was striking. The man's earthly remains were about to play a small part in an unfolding drama involving major museums, obsessive and sometimes eccentric scientists, and an array of amateur collectors. It is a story marked by evolving efforts to understand the human body in the language of race and human history. These efforts sometimes clashed, competed, and even overlapped in complex ways.

Leaving dark trails of blood, the soldiers dragged the corpse across the grass to a nearby fort. Word of the killing spread quickly. White civilians began gathering to celebrate. Settlers beat the lifeless body. Bones cracked. The scalp was cut off and carried away as a souvenir.[2] Once the settlers were finished, someone hastily buried the body in a shallow grave.[3]

In the days that followed, one German American newspaper reported on the skirmish from the settlers' perspective. The paper proclaimed, "It is time to hunt down these red beasts with iron pursuit."[4] Newspapers crowed about the small victory over the Native Americans, but the incident did not settle the tensions that had boiled over two years earlier in the Dakota War of 1862, a conflict resulting in numerous deaths on both sides.[5] In this context, the single violent encounter would not have stood out. The man's grave might have been soon forgotten. But only a few months after the Dakota man's death, his skeleton was quietly removed from the ground. The bones were brought to a military doctor stationed at the fort who carefully laid them out on a makeshift wooden operating table. The acting assistant surgeon, a measured and experienced man named Alfred Muller, lamented the circumstances surrounding the young Native American's death and mutilation. In a careful hand, he wrote a letter describing the body as having experienced "unnecessary ill treatment."[6]

Muller no doubt possessed his own vivid memories of violence between settlers and the American Indian tribes residing nearby. Just a few years earlier, he had received high praise for his treatment of wounded settlers following one particularly grisly attack.[7] Despite his firsthand experience with frontier violence, he found the recent beating of the corpse of the American Indian man utterly deplorable. But for Muller, what was done was done. With the bones now laid out before him, he delicately handled and examined each one, steadily writing his own detailed notes about the body. The smell of the body was different now, many days later, earthier. Bones were indeed badly cut and damaged in some places—however, many individual bones had been spared injury. Muller was fascinated. Despite his feelings about the treatment of the corpse, he did not rebury the body after his careful examination. Instead, he boxed the remains and sent them to Washington, DC, where the U.S. Army had recently opened a medical museum. Muller believed the skeleton might be useful to scientific endeavors described in a museum catalogue he had read.[8] Several weeks later, he sent a second parcel containing the man's missing hand, which Muller had managed to procure from some unnamed source in the name of science.[9]

The remains, which were eventually moved to the Smithsonian Institution, were swept into an expanding project to understand humanity through a changing kaleidoscope of ideas about the human body, race, and, increasingly, human origins and prehistory. Scientists, eager for evidence to support their ideas, organized spaces colloquially known as "bone rooms." In these spaces, they studied the bones in an effort to classify the races and develop an understanding of the deeper human past. They relied heavily on collectors of all kinds to gather specimens. Professionals and amateurs alike—influenced by a broad spectrum of ideas—began gathering and organizing human skeletons from around the world. Museums concerned with natural history, medicine, and anthropology—in their quest to solve riddles connected to race and human history—turned to human remains for answers.

Starting around the time of the Civil War and stretching deep into the twentieth century, gathering human skeletal remains was a common intellectual, cultural, and social pursuit. Though not limited to professional collectors, the practice centered primarily on an important, changing, and diverse network of scholars and scientists affiliated with a number of museums in the United States.[10] Donations allowed certain museum collections to grow rapidly in major cities across the country. Bones were sometimes sent to museums unsolicited. Others were gathered with more systematic intent—carefully removed from cemeteries or other archaeological sites. The gradual, piecemeal, and sometimes haphazard acquisition of human remains and subsequent attempts to draw important ideas from their study eventually developed into an outright competition to fill bone rooms with rare specimens. The specimens, it was believed, helped illustrate the ideas considered most important to the scientists at a particular historical moment. Fueled by competing desires and ideas, several major museums grew to dominate medical and physical anthropological collections, hoarding and attempting to fully comprehend the utility of the newly acquired bones on a nearly industrial scale. The desire for scientific collections and competing ideas about race and the history of humankind fueled the growth of bone collections, which outgrew storage areas and spilled into hallways and occasionally onto gallery floors in exhibitions. Physicians and anatomists who came of age during the Civil War were keenly aware of efforts to systematically utilize human skeletons for science. Some sought out the chance to get involved in the project themselves. Those involved disagreed on how best to categorize the races, care for the bodies, and understand them in the tapestry of human history—but they agreed on the inherent value of the project to establish and build bone rooms.[11]

After arriving in Washington, DC, the Dakota man's bones were placed on display in the Army Medical Museum (AMM), though details about any possible exhibit are murky. The skeleton was most likely used in the late nineteenth century to teach visitors about an emerging field called "comparative anatomy." The bones would have been identified as those of a Native American man, a Dakota stand-in for many tribes across the Americas—a lone and broken man intended to represent a unique and vanishing race. In some instances, bones were presumed to be similar enough to be simply interchangeable within racial categories; if the jaw was too broken or shattered for display, the museum could replace the broken or missing bone with another, similarly sized portion of a different Native American skeleton. For a few decades following the Civil War, medical students likely pored over bones still bearing cracks and cuts—testimony to harsh beatings. At some point, too, the bones were probably used to teach young medical students about the severe injuries they might encounter on modern battlefields. Packages accompanied by letters, many with stories like Alfred Muller's, arrived almost daily at the museum from around the American West and from expeditions around the globe.[12] Gradually, the significance of these collections would be reflected in their growing ubiquity in certain museums.

The original goal of the AMM was to collect examples of battlefield injuries. Soon after the museum opened, however, remains like those sent by Muller further encouraged the curators to pursue a new project in comparative racial anatomy, a long-standing scientific endeavor to classify human races on the basis of physical features and appearance. Several influential individuals working in museums now believed this project to be possible on a much larger scale than once imagined.[13] Medical doctors, anthropologists, and other scientists in the United States and Europe came to believe that perceived behavioral attributes of different peoples—such as intelligence and industriousness—could be directly correlated with physical characteristics, such as the size and shape of the skull. Some even believed that racial attributes could be measured and, indeed, ranked on a grand scale of humankind.[14] George A. Otis, who personally collected and measured hundreds of skulls for the Army Medical Museum, concluded simply, "The American Indians must be assigned a lower position on the human scale than has been believed heretofore."[15] His conclusions, though drawn from skewed measurements and based on faulty assumptions about the size of the brain cavity and its link to human intelligence, were nevertheless offered with the certainty that ample evidence was thought to afford. While not all scientists were as bold and direct in their racist conclusions, collecting, studying, and displaying nonwhite human remains largely supported the scientific (and pseudoscientific) racism that dominated the era. In many respects, the practice reinforced existing and emerging colonial power dynamics veiled as scientific and social progress. While a few indigenous individuals, tempted by money or other compensation, were complicit in excavating graves for the benefit of museum collections, most resisted the practice. Organized protests against scientific determinism resulting in the return and reburial of nonwhite skeletons, however, came about only much later in the twentieth century.

Eugenics, a proclaimed scientific field that is completely discredited today, was in the mid-nineteenth and early twentieth centuries considered a genuine contribution to the understanding of racial characteristics and human history. To better appreciate the influences of these and other ideas, especially as seen through the development of museum exhibitions, we must place them in greater historical context. Recent historians have been harshly critical of the legacy of the amateur collectors (perhaps better described as *looters*) and scientists who established these collections, but few books have fully explored the interconnected nature of individual institutional histories and human remains collections as a social and cultural practice. Nor has much been written on people's reactions to bone collections when they were placed on public exhibition. This is admittedly a difficult undertaking, as museum visitors left only shadowy impressions and few accounts of their reaction to exhibits from this era. Nevertheless, some newspapers, magazines, and books did record the impressions of museumgoers.

As scientific theories evolved, human remains were sought out with unprecedented zeal.[16] Some collections grew more rapidly than others, as a result of opportunistic collecting and uneven interest among museums for obtaining rare bodies and body parts. Despite this sporadic growth, the final decades of the nineteenth century and the early years of the twentieth century constituted the most active period of collecting human remains for museums—a legacy critical to those who are concerned with the modern-day fate of these mummies, medical specimens, and human skeletons. Significant ideas about the human body were hotly contested between the late nineteenth and early twentieth centuries, and scientists frequently turned to human remains collections for evidence to support new theories responding to old questions. These questions touched on many apparent problems and emerged in unique forms over time. Why do humans from distinct places appear different? What happens to our bodies when we age? Are some people inherently better suited to thrive in the natural and modern world, and if so, why? In transitioning from grave to museum bone room, human remains were endowed with new and powerful scientific meaning. By the turn of the century, skeletons became a key tool for testing the numerous theories surrounding race that were developing across a range of disciplines in the United States. At almost every turn, however, the grand vision laid out by the early founders of these collections—who claimed that secrets of racial evolution would be laid bare in the scientific examination of human bodies—seemed to veer further off course. Grave robbing, scientific racism, and ethnocentrism ultimately damaged the reputations of museums and scientists on a global scale.

As historian Steven Conn notes, "Museums functioned as the most widely accessible public fora to underscore a positivist, progressive and hierarchical view of the world, and they gave that view material form and scientific legitimacy."[17] Human skeletal collections were no different in this regard. But unlike material culture collections, they often had concrete and direct implications for theories about the medical and anthropological understanding of the body, and these ideas about race and human evolution ultimately permeated American culture. Ideas about the human body, particularly when related to race and gender, were critical in reinforcing—and sometimes deconstructing—basic cultural conceptions regarding humanity. Classifying the races implied that race was a viable concept and an accurate reflection of humankind. A binary mode of sexual distinction—classifying bones into male or female—was taken for granted. Human remains collections shaped significant perceptions about living and seemingly vanishing races around the globe. Indigenous bodies from Native Americans to Australian aborigines became especially prized. As scientists worked to articulate and defend their theories, they raced to build large bone collections to serve both as evidence and as explanatory models. Especially in the late nineteenth and early twentieth centuries, human remains were often presented on their own, lying silently under the glass cases lining museum halls and presented to the public as representing discrete facts with little added interpretation. Bones were simply points on hierarchies, stages in evolution, or illustrations of "comparative anatomy" and racial classification schemes.

Gradually, however, museums began to leverage bone collections—and artistic depictions of bodies drawn from the close observation of human remains—as elements of increasingly complex narratives about the past. The transition from an emphasis on discrete "facts" about race to a focus on more imaginative renderings depicting prehistory was neither clean nor wholly linear. In fact, the two concerns frequently overlapped in the minds of those who collected human remains. Embedded deep in many of the new narratives about human evolution or prehistory were descriptions of discoveries of the fossilized bones of human ancestors outside the Americas. Dominant narratives seemed to shift and change, depending on the intellectual arguments drawn out by curators and the dramatizations written by journalists describing exhibits. By arranging the evidence drawn from bones and fossils in a particular manner, museums presented visitors with visions of the human past seen through a changing array of ideas. This book explores how those changes—as messy as they sometimes were—shaped important and influential museum spaces as well as several major exhibits. Increasingly, an interest in human prehistory gradually displaced racial classification as the central scholarly debate consuming the obsessive drive to build bone rooms at museums of medicine, anthropology, and natural history in the United States. Despite this shift, the race question as it related to bone collections remained present, deeply rooted in their foundations.[18]

In the first two chapters, I cover the collection of human remains for science, particular discoveries that began to captivate broader audiences, attempts to document race through collecting skeletons in the early twentieth century, and important shifts connected to evolving ideas about racial classification theory. The next two chapters offer case studies: first examining the history of arguably the most popular and influential medical museum in the United States, the Mütter Museum in Philadelphia, and then exploring the early history of the San Diego Museum of Man, thought to be the largest exhibition about race and prehistory to date when it opened in 1915. The exhibit set an important precedent for blending artistic representations of ideas in anthropology with actual human skeletons and mummies. I then trace scientific racism's intense rise and fall in the United States as considered through the collection and organization of human remains in museum contexts. As the problems of racial classification theory were exposed, scholars studying and collecting human remains in North America began to shift their attention to a longer view of human history. The final chapter explores emerging studies about prehistory alongside a declining focus on racial classification in the

United States. This budding emphasis on prehistory stemmed from new discoveries and shifts in ideas about anthropology as a science both inside the museum setting and as a result of outside influences.

Key players in this drama include some of the most accomplished and controversial anthropologists of the time, including Franz Boas, Alfred Kroeber, and Earnest Hooten. The tradition established in the medical community—by physicians like Alfred Muller who were increasingly receptive to the idea that certain skeletons might be preserved for future research and teaching—resulted in the creation of a modest number of medical museums, found mostly in major urban centers like Philadelphia.

Central to this story is the Smithsonian Institution, which amassed the largest collection of human remains in the United States—if not the world—under the guidance of the fervent and divisive Czech-born American scientist Aleš Hrdlička. Following the Smithsonian's lead, museums in cities such as New York, Chicago, and San Diego collected human remains from around the world. Rivalries and disagreements over collecting information about the human body, including the intellectual and cultural ownership of remains, became increasingly heated in the early years of the twentieth century, only to die down as the intellectual center of physical anthropology moved away from the museum to universities. Figures like Hrdlička and Boas worked together as often as they stood in personal and professional tension, disagreeing over ideas pertaining to race and occasionally expressing distaste at each other's approach to science.

Despite the waxing and waning influence of these collections in American culture, issues surrounding the process of gathering, researching, and displaying human remains do not represent a simple declension story; debates surrounding human remains collections reemerged in new forms later in the twentieth century. Ethical challenges from indigenous communities—including demands for ancestors to be returned for permanent reburial—reshape the story. New controversies again gave museum bone rooms a prominent place in popular and scholarly discourse, though this contest was starkly different from that imagined by the founders and early organizers of these collections. An engine driving this story was the emerging competition over remains and clashing ideas over their exact meaning and significance. In 1901, for example, a collector working on behalf of the University of California, where a university museum was quickly being established, remarked on seeing a private collection of archaeological materials, "I obliged [the owner] to bequeath the whole to the State University, so that this might not go East or abroad."[19] Indeed, while museums frequently cooperated in their efforts to build skeletal collections, an underlying sense of competition compelled the drive for the best specimens, including discoveries of human remains thought to be particularly useful in teaching scholars about race or the past. This book considers how museums determined and reassessed the value of remains for both scholarship and exhibition.

The gradually deteriorating bones mostly languished on museum shelves, but the ideas surrounding them constantly evolved.[20] Once removed from the ground, taken from a battlefield, or excavated from other kinds of burial sites, bones often sat for generations behind the scenes, moved only periodically to be measured or displayed. Key perceptions related to these remains, however, changed drastically among the scientific community, the public, and indigenous communities around the globe between the mid-nineteenth century and the mid-twentieth century. This phenomenon was not confined to the United States. Human remains collections in European medical schools and museums reached similar sizes. Collecting and exhibiting bodies on such a vast scale in an effort to understand race, however, was unique to the United States. Exhibitions featuring human remains frequently offered explanations of the unique racial landscape in the modern United States; from that point, the project expanded out globally.

Historically, human remains may have received less scholarly attention than other kinds of museum collections, but the claims generated by their study rippled through numerous scientific and academic fields beginning in the mid- to late nineteenth century. Newspaper accounts of the exhibits helped capture the public imagination, fueling interest in bodies, race, and prehistory, connecting these subjects to skeletons or mummified bodies. With the decline of scientific racism in the interwar years, prehistory largely displaced racial classification as a focus of anthropology and archaeology. While the scientific theories that these collections supported stood on shifting ground, museum exhibitions nevertheless grew during the decades between the First and Second World Wars.

Other scholars writing about the history of archaeology, anthropology, museums, and collecting have largely focused on institutional narratives.[21] Later histories opened a rich field in museum history in the United States, but these too focus primarily on institutionally based material culture collections and displays.[22] Together, these histories argue that the study of material objects dominated anthropology in the United States until the 1920s or 1930s.[23] While these histories contribute mightily to our knowledge of the history of museum anthropology, they have largely left a documentary and interpretive gap surrounding the history of collecting, displaying, and studying the remains of human beings from around the globe. Within natural history museums, these collections were defined as belonging to physical anthropology. In more modestly sized medical museums, bones were studied for a variety of reasons; however, racial science and human history were subjects extending into medical circles as well, with collections of bones primarily understood as tools to advance the education of physicians, especially in the nineteenth century.

Scholars studying the collecting of human remains in the United States have tended to examine the practice within several contexts.[24] Related works on the history of physical anthropology, both inside and outside of the museum, have appeared in print since the 1970s. Central to the development of this literature was the historian Frank Spencer, who studied the emergence of a professional field of physical anthropology in both Europe and the United States.[25] Other historians and anthropologists examining the body have explored its materiality, focusing on the abstract and metaphoric meanings drawn by medical, anthropological, psychological, or other theoretical thinkers.[26] Parts of the body, or "specimens," experts argued, represented *facts* that could be placed together to showcase broader interpretations of race, disease, and history.[27] Scholars examining death and burial in the United States have created a diverse literature, examining the growth of cemeteries, the rise of the funeral industry, and the secular and spiritual meaning of death throughout the history of the United States.[28] Human remains collected for "science" were interpreted as having lost their spiritual context once taken from the cemetery, burial ground, or morgue. Once the body became an *object*, it became a tool in scientific study and display. This transformation led to a series of major ethical questions surrounding the treatment of human remains by museums and universities in the United States.[29] While the connection between scientists working with museum collections of human remains has been taken for granted in most works on the subject, a history of the growth of these collections and the ideas drawn from them during what was perhaps their most active and influential period has never been written.

Studying how museums collected, researched, and exhibited human remains is very different from examining the development of material culture collections in an effort to better understand museum anthropology. Museums building collections of human remains actively debated ideas as they chose what to collect, established their priorities for research, and determined how best to display ideas through exhibitions built around human skeletons. Museum anthropologists and medical doctors primarily concerned with human remains engaged with a different scholarly audience than did biologists and sociologists, and their discourse surrounding race and human prehistory embraced a trajectory decidedly different from that of the scholarship surrounding material culture. The arguments about race and human history were often made on the weight of physical evidence drawn from human remains (and gradually a small number of fossils and replicas). Moreover, the center of the field of physical anthropology remained in museum contexts until the Second World War, even as many studies on material culture shifted to university campuses.[30] Medical museums, which for a time collaborated with natural history museums, declined in significance for medical training largely because of technological developments in medicine rather than major theoretical shifts, as in anthropology. Medical museums, therefore, are at times critical to the story of collecting human remains, but they repeatedly fade in and out of the dominant narratives and discourses—including in recent debates concerning the ethical display of indigenous bodies surrounding mainly the collections of universities and natural history museums. Medical museums also witnessed a changing series of ethical guidelines for proper collecting and display of human remains; however, this debate has been less obviously situated in indigenous activism.

Rather than fully entering into the meaningful contemporary debates surrounding human remains collections, this book provides a more complete context for the history of bone rooms. Making the case for the cultural significance of these bone rooms is easier than gauging their exact size.[31] Remains are spread throughout large and small museums across the country, and cataloguing information is often vague and limited, though the information that museums provide to tribes, researchers, and casual visitors has grown much more detailed in recent years following the completion of federally mandated surveys. Recent estimates have placed the number of Native American remains in U.S. museums at about 500,000.[32] Adding to this figure are smaller collections of bones from African Americans, European Americans, and indigenous peoples from around the globe. It is estimated that museums in Europe have acquired an additional half a million sets of Native American remains since the nineteenth century.[33] More than 116,000 sets of human remains and nearly one million associated funerary objects are considered by museums in the United States to be culturally unaffiliated, meaning no specific ancestral origin has been ascribed to them.[34] Although potentially surprising to a museum visitor, these estimates of the size of human remains collections in the United States and Europe are conservative.

While modern ethical debates became increasingly vocal and articulate later in the twentieth century, the course of these complex issues was largely determined by the history of collecting and interpreting these remains before the Second World War. The history of these collections is dramatic, occasionally punctuated by unexpected twists. The story emerges from an ongoing competition to establish the largest and most prestigious museums in cities across the United States.[35] At times driven by both ego and intellect, scientists established a new field as they collected, their studies working to shape ideas about race and what it means to be human. For scientists who collected the dead, the desire to obtain remains for growing bone rooms often suspended or displaced codes of ethical behavior. Museum curators, as well as amateur collectors, competed and collaborated to understand the body as a scientific object; at the same time, visitors to museums that displayed bodies were continually enthralled, almost surprised, by the *humanity* of ancient and recent bodies they found exhibited before them.

CHAPTER 1

COLLECTING BODIES
for SCIENCE

Starting in the 1870s, whispers about fantastic discoveries in the American West began attracting attention. As settlers and explorers roamed the continent, some parties uncovered unusual and remarkably well-preserved human bodies previously unknown to science. Tucked deep into caves, buried near elaborate ruins, or frozen in arctic permafrost, naturally mummified corpses stupefied early observers. Unbeknownst to many of the people uncovering remains, the bodies were becoming critical in shaping emerging fields, including archaeology and anthropology in North America. At issue was not just what to do with ancient remains when found but also how best to interpret their significance. Despite the potential importance of mummies for understanding prehistoric history in the Americas, many people viewed these ancient corpses through the same lens driving the collection of skeletons of the more recent dead—scientific theories about *race*. When new bodies were discovered, the first questions usually focused on the *racial origin* of the mysterious bodies and their relationship to the modern races of the Americas. While the age of these mummies made them an important rarity, the primary purpose for collecting and permanently preserving new skeletons was to fit them into the puzzle that was the taxonomy of races. Nineteenth-century scientific publications and debate about racial divisions seemed only to fuel popular desires to collect human skeletons and mummies. Museums in the United States, aiming to catch up with their older European counterparts, began collecting bodies in North America with heretofore unseen zeal.

Naturally mummified remains fascinated scholars yet were also recognized as more than mere tools for archaeological science. In the late nineteenth century, American mummies captured popular imagination in a manner parallel to the prevailing Victorian-era obsession with Egypt in Britain. Certainly, the colonialist mania for collecting Egyptian antiquities took root in North America at the same time, but the puzzle of mummies from an ancient civilization in the American Southwest diverted attention to the mystery of the so-called cliff dwellers.

Following the Civil War, the number of museums collecting skeletons and mummies with the intention of studying racial science grew at a slow but steady pace. Major museums in the United States gradually built human remains collections around the expanding sciences of physical anthropology and comparative anatomy. Museum leaders came to view acquiring skeletons as a wise investment in emerging scientific disciplines. Natural history museums, in particular, began collecting skulls of nonwhite individuals, hoping to build on the work of people like the professor and physician Samuel George Morton, who had earlier amassed a collection of several hundred skulls and published a series of highly influential books based on his observations. Now considered inconsistent, pseudoscientific, ethically flawed, and marred by racism, Morton's work was nevertheless highly regarded at the time. The influence of his work helped frame future debates in physical anthropology.[1]

Like Morton, early museum collectors opportunistically collected skulls from distant contacts and acquaintances. Many also fashioned racial taxonomies after Morton's theories of the existence of particular racial groups. Private physicians and early amateur collectors, too, established skull collecting as a popular tradition in the United States. Mysterious packages would arrive at museums—sometimes accompanied by vague, handwritten notes with brief descriptions of the bones inside. Medical officers working as agents for the Army Medical Museum (AMM) were among the first to systematically collect skeletons for a major museum collection in the United States. Where army medical officers first trod, postal carriers soon followed, making it possible for amateur collectors to ship skeletons from far-flung locales to museums in Chicago, New York, or Washington, DC. As professional archaeologists began their work in the American West, amateurs chose to collect skulls and artifacts on their own. For many museums, knowing the racial origin of the specimen was enough to assign it a catalogue number and carefully place it into a drawer in a bone room.[2]

In 1875, when mummies were discovered in the Aleutian Islands, the *New York Times* referred to them as "a discovery of momentous interest to the scientific world."[3] Archaeologists, anthropologists, and even private corporations that had been collecting in Alaska for decades recognized the discovery as a useful human-interest story for promoting scientific work and collecting.[4] Naturally mummified remains of human corpses discovered in the dry, high-altitude caves of the American Southwest, the skin on their faces pulled tightly back by centuries of quiet preservation, made for especially captivating stories in the print media. Mummies discovered in the Southwest were often wrapped in blankets or adorned with other sacred burial goods. The hair of the mummies, either peeking out from the blankets wrapping their bodies or flowing down from their heads, ranged from dark black to various shades of red and blond—discolored after centuries of solitude in caves or other dry locations that allowed for the preservation of the body. Americans alternatively viewed the mummies with fascination, wonder, and revulsion. Newly scrutinized when collected and placed on public display, the ancient dead discovered in the American West were transformed from curiosity to antiquity *and* scientific specimen. Despite their apparent utility in understanding history, mummies discovered in North America in the late nineteenth century were most commonly viewed through the lens of the ultimate American qualifier: race. These naturally mummified corpses might have the ability to teach about ancient civilizations, to be sure, but most people who viewed the mummies in this era hoped they might reveal secrets about the racial origin of the modern American Indian. Mummies dramatically captured newspaper headlines and were the subject of important popular museum exhibitions, but their collection was merely one component of a larger project to collect and classify human bones from around the world.

The lengthy *New York Times* article following the discovery of a new group of mummies from Alaska details the condition of the remains and informs readers that the bodies were scheduled for shipment to Philadelphia, where they were prominently exhibited at the centennial fair the following year.[5] The display of the Alaskan mummies at the 1876 world's fair built on precedents of displaying human remains, but research into nonwhite, ancient, or otherwise unusual bodies was to assume a heretofore unseen centrality; developing alongside the public's burgeoning understanding of what was represented by human remains was the scientific community's organized inquiries into comparative racial anatomy and prehistory.

Presented as scientific commodities and tools for solving riddles connected to race and time, human remains briefly assumed great prominence in the American consciousness. Collecting, displaying, and researching human remains became professionalized in several different contexts. Bodies were a significant attraction to popular audiences attending fairs and private museums in the United States in the late nineteenth century. The public read about their discovery in newspapers and works of fiction, and they became eager to view them firsthand. As significant, however, were the growth and organization of collections of human remains in museums—institutions conducting significant amounts of scientific activity—during the same period. Between the 1870s and the end of the century, theories surrounding the notion of race and racial difference came to dominate studies of human remains in the United States. The Alaskan mummies on display in Philadelphia were no exception. These mummies were but a small part of a growing project to define humankind through systematically collecting rare corpses.

When it came to ancient history and, in particular, mummified remains, the Old World seemed to possess vastly superior and more significant relics (for some people, this was read as evidence that underscored notions of cultural and racial superiority). Attempting to satisfy a thirst for the valued relics of the Old World, Americans collected large quantities of both objects and remains through both professional archaeology and private patronage in Europe, Africa, and Asia.[6] As the nineteenth century came to a close, discoveries in the American West offered new spaces for professionals and amateurs to hunt for ancient remains. The remains found within or near cliff dwellings came to be defined as distinctly *American*—set apart from the ancient history of the Old World. Prevailing racism and popular notions of savage, static, and simple American Indians contradicted simultaneous efforts to preserve sophisticated ancient ruins, but most observers were either blind to or willing to accept this incongruity.

Only one year before the lauded mummy discovery by the Alaska Commercial Company, another *New York Times* piece articulated the popular antipathy toward the antiquities found in North America: "Our people are not much inclined to think of a great antiquity as belonging to the inhabitants of this continent, or to value highly the relics of our extinct human races. The popular contempt for the red Indian, and the knowledge that all which can be preserved of his tools, implements, and weapons, and works of art, form but a poor collection of antiquities, are in part the explanation of this indifference."[7] Running counter to this prevailing racial and cultural bias, however, were impressive new discoveries of previously unknown archaeological treasures. These finds gave rise to an almost frantic race to collect them for museums in the United States. Archaeologists and their sponsoring institutions engaged in political maneuvering and rapid collecting throughout the American Southwest in a complex competition for artifacts.[8] Museums sought ancient pottery and baskets from around the world, just as they sought ancient human bodies. Anthropologists, archaeologists, and physical anthropologists often viewed themselves as competing against both time and each other—but looters represented the main threat to building bone collections. Museums feared that the best specimens—those perceived as valuable for unlocking racial secrets through science—were rapidly vanishing. They were willing to take steps to protect them (at least so as to preserve them for future collecting by professionals). The race to collect human remains was on. In the course of the competition, museums variously became rivals, trading partners, and collaborators in efforts to fill bone rooms with remains from around the world.

A CRITICAL JUNCTURE

Events converging in 1879 critically shaped both archaeology and anthropology in the United States. Breakthroughs fueled continued growth in the desire to collect, study, and display bodies over the ensuing decades. Congress authorized the founding of the Bureau of Ethnology (later called the Bureau of American Ethnology, or BAE), and Frederic Putnam published a major volume on the archaeology and ethnology of the American Indians of the American West. At about the same time, Lewis Henry Morgan, then the nation's best-known anthropologist, was elected president of the American Association for the Advancement of Science (AAAS). Morgan's election to the head of a major scientific organization was a bellwether for anthropology's growing stature within the broader scientific community. During the same year, the Anthropological Society of Washington and the Archaeological Institute of America both came into existence.[9] The confluences marked 1879 as a critical juncture for professional anthropology in the United States.

One year following the election of Lewis Henry Morgan to head the AAAS, he appointed a Swiss explorer and writer named Adolph F. Bandelier to explore the ruins in the American Southwest. Before the expedition, Bandelier briefly visited John Wesley Powell, an internationally recognized scholar and explorer who was familiar with the region following his famed Colorado River expeditions. Powell, who lost an arm in the Civil War, encouraged Bandelier in their meeting to carefully examine the state of newly discovered archaeological monuments in the American West and report back to scientists in Washington, DC. By August 1880, Bandelier traveled west to begin a study of a series of ruins at Pecos, New Mexico. What he found at Pecos was unsettling. Describing his initial reaction to viewing the ancient dwellings, he wrote, "Most . . . was taken away, chipped into uncouth boxes, and sold, to be scattered everywhere. Not content with this, treasure hunters . . . have recklessly and ruthlessly disturbed the abodes of the dead."[10] The bones, mummified flesh, and burial goods, naturally preserved over many centuries, now faced haphazard theft and abusive destruction.

The newly founded Archaeological Institute of America expressed concern to officials in Washington, DC, about ancient sites, but such concern resulted in little action. Many people within the federal government simply cast doubt on the idea that all such sites could be protected under federal law.[11] Despite a slow start, newly formed professional organizations concerned with promoting archaeological research and protecting antiquities proved to be significant in the movement to preserve archaeological sites in the United States. Central to the growing concern over preserving antiquities was the desire to collect the bodies for science.

By the time the AMM began collecting bodies for studies in comparative anatomy, Samuel George Morton had already amassed sizable personal collections in Philadelphia.[12] Morton's skull collection, obtained through contacts spread around the globe, provided a clear example on which the physicians at the AMM could base their own collections and research.[13] Indeed, when the AMM published a catalogue of crania collections, the work was modeled after Morton's widely read book *Crania Americana*. The AMM hoped to expand on this research by making its own catalogue available to an even larger number of students and scholars studying comparative racial anatomy around the country. Buttressing ideas with detailed measurements of all aspects of the human skull, scholars were discussing race in what appeared to be an increasingly complex and sophisticated manner.[14] The combined work of scholars like Morton and those at the AMM laid the foundation for the rapid expansion of museum collections of human remains in the United States.

Morton was a leading figure in American science during his lifetime. Considered by some scholars to be the intellectual father of physical anthropology, his human crania collections, by 1849, contained the skulls of eight hundred individuals.[15] The skulls varied in age, completeness, and origin. Some were stained white due to prolonged exposure in the hot sun, others a deep mahogany brown from the earth they rested in before being disinterred. Morton's collection, initially organized for pedagogical purposes, eventually enabled him to produce studies supporting the idea that the measurement of cranial capacities helped identify particular races. Each skull, upon its acquisition, was carefully measured, labeled, and delicately placed on a shelf for preservation. In the words of archaeologist David Hurst Thomas, "To Morton, the human skull provided a highway back in time, a way to trace racial differences to their beginning."[16] For Morton, races were unchanging and arose as a result of distinct and separate origins, a concept known as polygenesis.[17] Others argued in favor a competing idea: monogenesis, or the notion that races arose from a single lineage (often starting with the Judeo-Christian idea of Adam and Eve in the Garden of Eden). This single creation of humankind, monogenesists believed, continued evolving into distinct races. Others, like Morton, argued that direct observation of human skulls proved otherwise.[18]

Morton made numerous conclusions based on his collection, including that Native Americans were a race distinct from the Inuit and Mongolians. Intelligence, or more specifically cognition, was thought to be determinable through measurements taken from the shape and size of the skull. Morton eventually stretched his conclusions, based largely on these measurements, to argue that races actually represented distinct species.[19] In 1981, the biologist Stephen J. Gould reexamined Morton's collection and argued that Morton had intentionally distorted data in order to present Caucasians as an intellectually superior race. Gould presents Morton's conclusions alongside numerous other examples of scientists attempting to reach similar conclusions about race-based intelligence, namely, that whites were naturally more intelligent than blacks or American Indians.[20] Gould's high-profile work was part of a new wave of scholars who continued to deconstruct racial classification theories, describing them as distinctly pseudoscientific and racist. Some in the anthropological community have since challenged Gould's argument that Morton *intentionally* distorted measurements, while other scholars agree the data was skewed on purpose.[21] Regardless of intent, Morton read his evidence as directly proving the existence of distinct racial groups. Certainly,

American and European intellectuals had been keen on the idea of scientifically classifying the races for some time, but Morton's presentation of detailed skull measurements alongside striking illustrations projected scientific rigor, limited only by tantalizingly incomplete data. Morton's work seemed only to beg the collecting of more evidence.[22]

Morton hesitated before stretching his argument to present whites as a superior race, yet he did little to prevent his readers from making the small mental leap to that conclusion.[23] While Germans, English, and Anglo-Americans possessed the largest cranial capacities, the smallest belonged to American-born blacks. Native Americans were described by Morton as "averse to cultivation, and slow in acquiring knowledge; restless, revengeful, and fond of war, and wholly destitute of maritime adventure."[24] He drew these conclusions from the sizes and shapes of the skulls of other races, arguing that the smaller brain cavity of Native Americans led to a decreased average intelligence. Upon his death in 1851, Morton was eulogized in the antebellum South as having helped to definitively and scientifically prove the inferiority of the African in relation to the European.[25] The evidence presented by Morton lent itself to obvious scientific and social conclusions, mostly supporting the status quo of race relations in the United States in the middle of the nineteenth century. Although Morton's publications were widely read during his lifetime, his influence waned following his death. This was partially because he had few students.[26] Morton's work also became increasingly associated with an anti-Darwinian, pro-Confederacy alignment unpopular after the Civil War. In particular, French anthropologists, who were largely opposed to the institution of slavery, worked to separate themselves from Morton.[27]

Despite Morton's gross inaccuracies and distortions, his studies are especially significant in that they were based heavily on the close observation and study of collections of human remains gathered from around the world. His work underscored the growing perception behind the utility of skull collections for future science. Morton's work helped to create a professional precedent, signaling to young scholars that the question of race could be understood through the collection and study of human skeletons. During his own lifetime, Morton's work was critical to the development of medical museum collections like the AMM. While his standing diminished in later years, collecting, measuring, and studying the skeletons of races from around the world continued and even expanded as a practice. Racial classification and racism appeared inherently and inseparably linked.

THE ARMY MEDICAL MUSEUM

Established by Surgeon General William Hammond in 1862, the AMM was initially created for the purpose of collecting examples of battlefield pathology during the Civil War.[28] Indeed, even after the war, medical officers complied with orders to collect bodies and body parts that illustrated wounds, diseases, or the result of surgical procedures encountered by physicians working in the field. As the museum developed, the interest of the curatorial staff evolved into a burgeoning area of science that they called simply "comparative anatomy." For museum curators in the era, the broad influence of thinkers and collectors ranging from Samuel George Morton to Charles Darwin and figures like Thomas Jefferson (who collected and wrote about the skeletons of Native Americans in Virginia) proved foundational, opening new possibilities for the intersection of studying race and collecting human skeletons. These and other thinkers, in using human skeletal remains as scientific (or pseudoscientific) evidence, pushed museums to apply techniques of comparative anatomy across humankind.[29] These ideas gradually extended into museum halls, where battlefield wounds and birth defects were examined alongside the display of skeletons of American Indians, reptiles, and birds.[30]

The AMM was divided into several sections, including a large "anatomical section," focusing on collecting "normal" human skeletal material. The anatomical section, in the era following the Civil War, focused mainly on human skulls. Most human remains collected for the museum for the purposes of comparing the anatomy of different races were from American Indians, though remains from humans of Europe, Africa, Oceania, and Asia gradually complemented these collections. While the AMM's interest in comparative anatomy pushed it to collect bodies from human populations around the globe, the remains of indigenous groups were emphasized. When opportunities arose, the museum collected the remains of African Americans and a smattering of European Americans from North America.

In 1866, when the museum moved to the recently vacated Ford's Theater, the site of Abraham Lincoln's assassination only one year earlier, it presented a somber tone on the top floor of the building. With the site open to the public, museum leaders assumed that most visitors would recoil at seeing medical specimens in glass jars and the skeletons of victims of violence. Nevertheless, in spite of the new location, or possibly because of the traumatic memory associated with the building, visitors began pouring into the museum. One army physician argued that the location of the museum was fitting, eerily invoking the assassinated president: "What nobler monument could the nation erect to his memory, than this somber treasure house, devoted to the study of disease and injury, mutilation and death?"[31] Slowly, the museum became one of the most popular tourist destinations in Washington, DC. Although the museum was open for only four hours a day and children were not allowed admittance, it received about six thousand visitors in 1867. Within a matter of only four years, the number of visitors to the small museum tripled. The first curator of the museum, Brigade Surgeon John Hill Brinton, observed, "The public came to see the bones, attracted by a new sensation."[32] Despite the astounding popularity of such exhibits, museum leaders maintained inaccurate assumptions about popular audiences and what they would deem repulsive or compelling.

Several decades following the founding of the AMM, the army and the Smithsonian agreed to transfer ethnographic materials acquired by the AMM to the Smithsonian. The Smithsonian, in turn, promised to exchange human remains relevant to the army's research to the AMM. This was no idle promise. In fact, the Smithsonian and the Bureau of American Ethnology (BAE) soon began mounting expeditions across North America that turned up human skeletons, including John Wesley Powell's famed survey of the Rocky Mountains. These federally sponsored expeditions routinely discovered and collected human remains, many of which were subsequently turned over to the AMM. Although the AMM received a highly visible portion of its collections from such publicized scientific expeditions, the bulk of the collections in the comparative anatomy section quietly arrived in boxes shipped from army and navy medical officers in the American West and around the globe.

In the early years of the Civil War, the surgeon general authored a circular calling for medical officers to collect materials of "morbid anatomy, surgical and foreign bodies removed, and such other matters as may prove interest in the study of military medicine or surgery." He added, "These objects should be accompanied by explanatory notes."[33] In 1867, the surgeon general again wrote a circular, instructing medical officers to "collect crania together with the specimens of Indian weapons, dress, implements, diet, and medicines." This collection, the circular explained, intended "to aid the progress of anthropological science by obtaining measurements of a large number of skulls of the aboriginal races of North America." It was necessary, therefore, "to procure sufficiently large series of adult crania of the principal Indian tribes to furnish accurate average measurements."[34] Building on the tradition of Morton, the medical museum wanted to aid the young science of anthropology by collecting thousands of skulls and skeletons.

Word that the AMM was interested in acquiring bones spread throughout military ranks, and personnel around the world obliged. Skeletal material began arriving in Washington almost daily. The army's official call for specimens echoed throughout the American West, reverberating as well in military expeditions elsewhere in the United States and naval stations around the world. Following the distribution of the circular, shipments began appearing from around the globe. The purpose of the collection, early on, was to further the professional education of medical officers through the study of various medical conditions potentially encountered during military service. The assumption of the early circulars was that medical officers in the U.S. military might benefit from two kinds of collections, those demonstrating the various forms of battlefield injuries common during wartime and those of the exotic bodies of nonwhite races surrounding the military bases of the U.S. Army.[35] The letters accompanying parcels to Washington were at times matter-of-fact about the shipment of human remains to the museum. Occasionally, however, letters sent to the museum contained vivid, even disturbing details. Stories of collecting human remains sporadically contained sentiments of excitement and feelings of danger. Gathering bones, as some people described it, was a nearly unparalleled adrenaline rush. Removing skeletons from sacred graves was understandably an affront to most Native American tribes, and some stood ready to protect their ancestors with force. Officers working on behalf of the medical museum were, ironically enough, willing to risk life and limb to collect bodies or parts of the dead.

As the letters and boxes poured into the museum, curators may have been shocked or mystified by some stories scrawled into the correspondence accompanying bone shipments. In 1867, a medical officer named W. H. Forwood wrote a letter to accompany the shipment of a human skull. Forwood was stationed at Fort Riley in Kansas, but he had acquired a skull that he claimed belonged to a Cheyenne killed at the Sand Creek Massacre in 1864. The letter fails to articulate how, exactly, Forwood acquired the skull some 350 miles from Sand Creek.[36] In fact, numerous AMM acquisitions were reported to have been collected from the site of the Sand Creek Massacre.[37] Bones were also collected from Little Big Horn, as well as the site of many smaller skirmishes and massacres.[38] Remains were acquired from burial sites as often as they were from battlefields. Because American Indians confined to reservations were commonly buried at or around army forts, medical officers could easily obtain their remains.[39]

Medical officers, in fact, took advantage of nearly every opportunity to acquire the remains of nonwhite individuals for the AMM. This occasionally included African Americans. Remains from Vicksburg, Mississippi, representing eight African American individuals arrived at the museum in 1869. The collection included the remains of a black infant, which the AMM considered especially important.[40] Scientists prized infant skeletons for two reasons. First, they represented early developmental stages in normal human anatomy. Second, due to their fragile and small nature, they were simply harder to find.

The AMM was interested in acquiring both white and nonwhite remains, but the number of American Indian and African American bodies that the museum acquired vastly outpaced the number of European American remains collected by the institution. The skeletal materials of white individuals that *were* acquired for the museum's collections often included those relegated to the fringes of society—criminals, the destitute, and those with unclaimed remains.[41]

Despite the apparent demise of phrenology—a study whose proponents argued that certain functions could be directly mapped onto the size and shape of the brain and skull—comparable ideas continually resurfaced in racial classification theory over the next several decades. In May 1872, the AMM received a shipment that demonstrated how earlier anatomical theorists continued to subtly influence collecting human remains. A letter accompanying a pair of crania, written by medical officers working for the Japanese government, explained the origins of the two skulls. The first was a cranium of what was described as an "educated Japanese gentleman," and the second was the cranium of a "Japanese criminal." The men were close in age at their respective deaths, the "gentleman" dying at about age thirty-five and the "criminal" thirty-four. The underlying presumption in the donation was that the nature of their education and their lived experiences may be reflected in the size and shape of their crania, delineating specific ranges within the supposed races of humanity. Japan, like many regions in the United States, outlawed most medical dissections, and therefore human remains or skeletal materials proved difficult to acquire.[42]

An undated letter, written from Fort C. F. Smith in Montana by a medical officer named James P. Kimball, exemplifies the opportunistic nature of military skull collecting. The letter accompanying the shipment describes the enclosed materials as "three . . . Indian Crania—Blackfeet—Picked up on the Rose Bud Creek . . . at a place said to be the site of a former battlefield between the Crows and the Blackfeet in which . . . the Crows being the victors, carried away their dead leaving the bodies of their enemies upon the field." The assistant surgeon assured the museum, "The story is as well authenticated as any of the local traditions and it is considered of beyond doubt that these Crania are those of fallen Blackfeet."[43]

The memoirs of William Henry Corbusier (1844–1930) further illustrate how opportunistic collecting became for army medical officers. Corbusier was an army surgeon and prolific writer, recording his experiences in the American West and later in the Philippines

In February 1875, he was stationed among the Yavapai when he was ordered to help supervise the displacement of fourteen hundred Indians from the Rio-Verde Agency. The group was ordered to walk about 150 miles through rough terrain. Conditions were harsh, and the group grew tired and hungry, having been forced to leave behind the foot-sore cattle intended to serve as food. Eventually, a group of Native Americans turned on the military officers, and the unarmed Indians were driven back by the officers, including Corbusier. Corbusier's account reads, "I collected ten wounded men, whose wounds I dressed, and found four dead, shot through the head. These were buried and on my way back to Camp Verde, I disinterred the heads and sent the skulls to the Army Medical Museum, as they showed the so-called explosive action of a bullet passing through the skull which it broke into many pieces."[44] Corbusier's account might be surprising given the paternalistic sympathy he shows for the Yavapai throughout the memoir.[45] Nevertheless, his decision to collect the skulls of the deceased American Indians and send them to the AMM was common, and it demonstrates the overwhelming influence of the scientific orders circulated among army surgeons stationed throughout the American West.

The AMM also worked with medical colleges to collect the remains of nonwhite individuals. In 1869, the skull of an elderly African American male arrived at the museum via Georgetown College, the body previously having been used for dissection.[46] Georgetown later contributed the remains of other individuals similarly used in teaching anatomy through dissection. Medical students, still responsible for acquiring their own cadavers for dissection, sometimes even sold to the AMM the crania of the cadavers used in their studies.[47] Though these collected remains include the bones of at least one white female, most remains acquired following dissection were of nonwhite or mixed-race individuals.[48]

AMM records indicate that military medical officers took a great personal interest in the project to establish and build human remains collections. Although the surgeon general's circular was considered an order, remains were clearly collected out of both personal and professional interest. In 1879, the AMM received a letter from a surgeon who claimed to have collected a cranium a decade earlier, shortly after receiving a copy of the circular order. The surgeon wrote of the circular, "I complied with that order to the utmost of my ability. I then forwarded the collection to Washington as ordered except one Ogalalla skull of a young squaw that died of phthisis. I retained that skull on account of the remarkable beautiful teeth she had—every tooth was perfect and of the most symmetrical order." He continued, "I secured the skull from a scaffold that was created on a high hill, over looking a small indian [sic] village. . . . As I got it in the day time, and before the eyes of many Indians, who could see me in the distance, I had a lively adventure with it." He concluded, "perhaps partly on that account, I held on to it as long as I did, as a trophy."[49]

As the surgeon general's call for specimens was published and republished, boxes continued to arrive in Washington. Gradually, the army surgeons spread around the American West acquired a common understanding of the "scientific" project of race-centered comparative anatomy. Although bones sometimes arrived with tantalizingly few clues as to their exact origins, they sometimes included intricate details about the deceased. Shipments might include people's race, tribe, sex, and age but also sometimes their name and a general portrait of their social standing within their community.[50]

Remains brought to the AMM for the purposes of comparative anatomy were understood to help advance theories about race. The project went in different directions over time. George A. Otis, who worked for the AMM between 1864 and 1881, worked to build collections intended to represent surgical methods and the comparative racial characteristics of humankind. Otis was particularly interested in the acquisition of skulls, hoping other scholars might utilize the crania for comparative racial studies, though the vast majority of his own work was focused on surgical techniques. Otis, working with other medical officers, published massive volumes recounting the medical history of the Civil War before issuing, over the course of the following decades, equally long and serious volumes detailing the advancements of surgical techniques.[51] Bone collecting had become a booming enterprise.

A NATIONAL COLLECTION OF SKELETONS FOR A NATIONAL MUSEUM

In 1897, William Henry Holmes, an archaeologist recently arrived from Chicago, visited the AMM. Holmes, now working for the Bureau of American Ethnology, eventually rose to a leadership role in the newly formed Department of Anthropology at the Smithsonian. Holmes was bearded and serious looking, but he maintained a steady and generally affable personality. People he worked with tended to be loyal to him. Trained as an artist, Holmes was renowned for his ability to capture geological formations in understated watercolor paintings. During the ensuing decades, he became a critical force for anthropology from within the Smithsonian.[52] As he toured the AMM, curators explained to him that their interests had shifted away from comparative racial anatomy. The shift caused thousands of human remains to now sit, unstudied, in wooden bone room cabinets. Eventually, Holmes requested that the collection be transferred permanently from the army to the Smithsonian, where he was confident the collection would advance the understanding of the races of humankind.

As a museum leader, Holmes possessed a unique ability to garner support for other projects and scholars. He also maintained impressive foresight, anticipating developments in the field and leveraging them to the advantage of the Smithsonian and the Bureau of American Ethnology. His decision to acquire the comparative anatomy collection from the army was especially significant in that it served as the impetus for creating a Division of Physical Anthropology within the U.S. National Museum, a central location for ideas and debate about the human body over the next century. The army quickly agreed to turn over its comparative anatomy collection and the nearly three thousand bodies, including the remains of the lone Dakota killed in 1864, transferring each of them across Washington, DC, to the Smithsonian. Although the skeletons themselves remained largely unchanged from when they were discovered, prepared, or removed from the ground, a sea change in both scientific and social interpretation surrounding these remains was starting. Their transfer indicated that it would be the field of anthropology, not medicine, to take the leadership role in espousing academic ideas about race. Many of these ideas were drawn directly from the study of human skeletal remains in the United States.

Of the 3,761 sets of remains ultimately moved to the Smithsonian, the vast majority were crania collected without any postcranial remains (meaning all parts of the body other than the skull).[53] Crania, in fact, represent nearly 80 percent of the collection.[54] Given the conspicuous examples of battlefield collecting, one might expect the AMM collections to be skewed in terms of sex. In fact, the numbers of individuals identified as male and as female are strikingly even.[55] This reflects the widespread looting and grave robbing across the American West, compounded with the relative sex balance of many massacre sites where the U.S. Army collected remains for the AMM, sites where both men and women were killed. A large component of collections originated from the American West: the collection includes 642 remains from California, 163 from the Dakotas, and 57 from Montana. Though collections mainly arrived from western territories, numerous remains did come from states like New York and Virginia, reflecting the wide range of locations in which army medical officers were stationed. Surgeons stationed at lonely outposts were also likely to have ample time on their hands, opportunistically acquiring remains as a side project. Just as the symbol of the sun-bleached skull came to represent the harshness of the American West, scholars in the eastern United States were hoping to utilize it as a symbol for understanding the secrets of racial difference. As behind-the-scenes museum collecting expanded during this period, exhibition of human remains also became more commonplace.

POPULAR DISPLAY OF "SCIENTIFIC" REMAINS IN GILDED AGE AMERICA

With bone collecting becoming an expanding practice in science, private entrepreneurs and showmen occasionally drew on scientific and pseudoscientific ideas to display human remains in different contexts. Ancient history, portrayed as mysterious, dramatic, and violent, was often used as a hook to grab broader audiences already visiting fairs and expositions. Audiences seemingly craved the sight of mummified bodies, grabbing headlines and drawing the biggest crowds. Although it would be a mistake to classify this popular interest as simple morbid curiosity, those who wrote about these exhibits continually noted the macabre sense of awe and wonder present in audiences touring fairgrounds and galleries. Although commercial displays of human remains were popular in the late nineteenth century, private collecting for exhibits, especially at world's fairs, began to fade after the first decade of the twentieth century. In this context, it is likely that some popularity transferred to museums that invited audiences in to view similar human remains presented in a subtly different light over ensuing decades.

Not all plans to create grandiose exhibits of human remains turned into profitable commercial ventures. In 1890, just one year before P. T. Barnum's death, the famed circus owner and showman proposed an exhibition for the World's Columbian Exposition that would surpass the displays of any previous world's fair. Barnum proposed to purchase, for the massive sum of $1 million, the mummified remains of the Egyptian pharaoh Rameses II along with the remains of much of his family. Rameses, as Barnum notes in his proposal, was a villain of the Old Testament, and therefore an untold number of Americans would be eager to view his earthly remains. In an article appearing originally in the *North American Review* and subsequently syndicated in newspapers around the country, Barnum wrote,

> Think of the stupendosness [*sic*] of the incongruity! To exhibit to the people of the nineteenth century, in a country not discovered until 2,000 or 3,000 years after his death, the corpse of the King of whom we have the earliest record! Consider, too, that that corpse is so perfectly preserved after thousands of years in the tomb that its features are almost perfect; so that every man, woman, and child who looks upon the mummy may know the countenance of the despot who exerted so great an influence upon the history of the world. And it might be a useful thought to this generation, proud of its scientific and mechanical triumphs, to bear in mind that the art that embalmed the body of Rameses so perfectly is lost, with a great many others that were known to remote antiquity.[56]

For Barnum, exhibiting human remains balanced on delicate interplays of showmanship, exploitation, and education. In Barnum's attempted purchase of the remains of an ancient king, which he intended to display for profit, he seems to recognize the desire Americans possessed for viewing recognizable historical relics, as well as the burgeoning desire to gaze on the foreign body of ancient history. Despite his grand plans, his proposal never came to fruition. Barnum represents probably the best-known proponent of this type of commercial venture, but other proposals focusing on the large and small exhibition of human skeletons and mummified remains imagined scenarios by which to profit from the exhibition of these bodies. Other commercial schemes to collect and display the bodies of the dead at world's fairs, however, actually proved successful, attracting the attention of museums that were soon to purchase many of these remains. By the turn of the century, mummified remains became a valuable commodity for scientific as well as commercial ambitions.

MUMMIES IN AMERICA

Human remains were not solely within the purview of medicine or anthropology. Nor were scientists, still heavily invested in comparative racial studies, limiting their research to corpses of the recently dead. Americans had avidly collected mummies from Egypt for museums for decades, but discovering naturally mummified bodies in the United States caused a new sensation. American archaeologists had yet to penetrate the emerging archaeological discipline into the Middle East, but colonial treasures making their way to museums in London and Paris were well known in North America.[57] Mummies, by this time, were already considered synonymous with pharaohs and pyramids. The notion that bodies of exotic and distant races had been mummified naturally in the desolate American West encouraged fantasies about unknown civilizations. Onto a blank slate of an unknown civilization, archaeologists and looters etched new ideas through massive collections of distinct pottery and mummified remains. Many early proponents of the antiquities in the region proposed wild theories disconnecting the contemporary American Indians from their ancestors in the Americas, claiming that only an outside civilization would be capable of constructing such grand structures as the cliff dwellings. This pattern of popular and academic disconnect, featuring rampant contempt for the modern Native Americans and a strange attraction to the mysteries of their apparent ancestors, became firmly embedded throughout the Americas.

Not only were boundaries stretched in terms of the ideas connected to mummified bodies, but also these newly discovered remains emerged from a region with which Americans were growing more familiar: the American Southwest. Despite an emerging popular interest in the region, many writers, artists, and scientists advanced the claim that American Indians generally lived *outside* history as it was commonly understood. Many took for granted the notion that Native Americans maintained primitive cultures in stasis for centuries. The modern Native Americans, many argued, never advanced beyond the level of Stone Age societies of the Old World, remaining immune to progress. Many gave little credit to the prehistoric relics of the Americas aside from the occasional nod to the Mayas and Incas, most often describing prehistoric discoveries as "rudimentary" or "crude." The study of the indigenous history of North America, therefore, was more or less an unknown at the beginning of the nineteenth century. As the century progressed, however, the issue of how to study the history of Native Americans became a significant component of a discourse surrounding the practice of history in the United States.[58] Especially critical was a growing awareness surrounding the mystery of the cliff dwellers. New discoveries revealing apparently complex civilizations—as well as mysterious burials including mummified corpses—brought forth new and complicated questions. Who were these people? Where did they come from? And most critically, how might they fit into the complex racial puzzle in the Americas?

The popular media soon joined archaeologists in pondering the cliff dwellers. Hundreds of articles appeared in newspapers around the country, reporting magnificent relics and forcing readers to reevaluate long-held notions about a primitive and unchanging indigenous culture. Many simply could not accept that the seemingly "primitive" indigenous peoples in their midst were capable of constructing grand structures or artworks. These articles reflected and fostered the nation's genuine fascination with the discoveries and also presented a newfound boosterism coming from the cities and towns near the ruins in the American Southwest.[59]

To a striking degree, reporting on the mummified remains found in the American Southwest captured both the popular and scholarly imagination. An 1887 article in *Harper's Magazine* reflects the widespread fascination, comparing newly discovered mummies to those from the Old World:

> There were recently lying in San Francisco, awaiting the shipment to Europe, the remains of four Arizona Indians, which are, perhaps, the most perfect specimens of the natural embalming process of a dry climate ever found in this country. These remains are simply dried up by the action of an atmosphere in which there is no humidity. Even the viscera, which all embalmers in Egypt found it necessary to remove in order to guard against decomposition, have been desiccated like the other parts of the body, so that one has here the practical result of the embalmer's art with not a single organ of the body removed.[60]

Despite discovering hundreds of naturally mummified bodies, scholars remained confounded by their origin.[61]

Mummies had already assumed a place in the American consciousness, but a major discovery led to an unforeseen level of prominence. In December 1888, Richard Wetherill and his companion Charlie Mason accidentally discovered ancient and elaborate structures built into natural rock formations in what is now Mesa Verde National Park in Colorado—stumbling across an ancient structure later called Cliff Palace. At the site, the two men uncovered a stone axe, pottery bowls, mugs, large water jars, and three human skeletons. Much to their surprise, these discoveries generated little interest from the public until they announced finding a naturally mummified child at another nearby site, Mancos Canyon. Human skeletons, though not exactly common throughout the American West, simply did not capture the public's attention to the same extent that a mummified body did.

After the ancient mummified child was found at Mancos Canyon, debate ensued in both scholarly and popular outlets. Newspaper articles demonstrated a lingering confusion as columnists desperately sought to definitively ascribe a *race* to the mummies. Writers and audiences struggled to understand a human body, even an ancient and mummified one, without a specified race. Race was so critical a qualifier, in fact, that without a discussion of it, an article about these mummified remains would have been sorely out of place. One typical article, published in an 1891 Grand Junction, Colorado, newspaper read, "They were neither Indians nor Esquimaux. They were not Negroes neither were they Malays nor Mongolians. All indications suggest that they were a white race. They had very soft hair in all cases. In some specimens it was very dark in color; in others reddish-brown, red and light blond." Scientists later concluded that the hair color might have changed due to a variety of factors, including gradual dehydration or exposure to sunlight. At the time, however, writers were so eager to connect the new discoveries to observable racial characteristics that they were willing to make astounding claims based on only a few small, and potentially misleading, clues. The same article continued by explaining the proposal to purchase the collections for the state, adding, "It is to be hoped that our legislature will appreciate the study of this extinct race enough to save this collection for the state." The author of the article argues that while some people in the region postulated a relationship between the cliff dwellers and the ancient Indians known as the Moundbuilders to the east, a simple comparison of the skulls demonstrates the racial difference between the two groups.[62]

Academic circles responded to the mystery of the cliff dwellers with growing interest, especially as the spectacular artifacts and mummified remains started trickling into museums and universities. In 1890s, a Smithsonian curator with whom the Wetherill brothers were corresponding, explained, "It has always been a source of regret" that officials in Washington dedicated such little time to protecting and exploring the archaeological ruins of the American Southwest.[63] Wetherill responded with a letter further detailing the vastness of the collection they had brought together: "skeletons and dried bodies from the smallest child to the full grown man, and skulls, from a number of which the bodies have decayed."[64] Wetherill certainly had a stake in promoting the value of the collection, and his emphasis on the number and type of human remains in the collection is additional evidence of their perceived monetary value. Archaeologists and amateurs across the American West became increasingly confident as the remnants of both bodies and cultures were sought-after objects for museum collections. Unfortunately for amateur collectors hoping to strike it rich, museums were generally too poorly funded to even consider paying the exorbitant sums initially demanded by some of the collectors. As the media publicized stories of the finds of ancient ruins, would-be explorers and archaeological profiteers wrote to museums in the hope that their travels would be backed financially by museums that wanted to add to their collections from the ancient sites of the American Southwest.[65]

Exhibiting these discoveries was a widespread practice and attracted audiences to locations outside of large, urban natural history museums. When a Forest Service employee accidentally discovered a small mummified body in Arizona, the body was sent to a nearby town and displayed in a drug store's window "for a couple of days" before being sent to the Smithsonian, "in order to give the public an opportunity of seeing it."⁶⁶ Just as with archaeologists and regional explorers who discovered remains, local cities and towns in the West were often given an opportunity to publicly view remains before their submission to museums in distant urban centers. Newspapers in Durango, Colorado, reported in 1892 that a collection belonging to a local man had gone on display and that the free, temporary exhibits in the town might represent the finest collection of ancient relics from the American Southwest anywhere in the country. One paper boasted, "It is questionable, indeed, wether [sic] the Smithsonian institute in Washington possesses so complete and varied a collection of relics of an extinct race." The temporary displays in Durango included a room featuring "ten mummified bodies and eighteen or more skulls, some with hair on them in a good state of preservation." The paper assured readers that the bodies were not merely on display as an appeal to the macabre. Instead, it argued, they "afford abundant food for study and investigation."⁶⁷

Some amateur collectors, if they were not inclined to desperately try to sell artifacts, were willing to part with their findings in order to deposit them at a noteworthy museum. Although the national craze for American mummies lasted only a few decades, its effect was significant in pushing museums to acquire mummies and skeletons. The national interest connected to discovering mummies in the Americas also pushed policy makers and commentators to define ancient bodies as valuable "antiquities" and objects for studying racial classification. Although the terms and definitions changed drastically over the ensuing generations, the cultural value of these discoveries, both accidental and intentional, had been firmly established.

The discovery of the spectacular cliff dwellings at Mesa Verde by the Wetherill ranchers in the 1880s and 1890s was one of the most important events in America's growing fascination with the prehistory of the American Southwest. Although the human remains became critical tools for understanding prehistory, their initial unearthing was couched in terms of the value for racial science. The decision to remove artifacts from Mesa Verde later became significant in the development of both state and federal protections for historic sites.[68] Before that, however, archaeological objects and human remains left behind by the ancient Pueblo peoples fascinated the American public in displays at museums and fairs.

HUMAN REMAINS AT THE WORLD'S COLUMBIAN EXPOSITION, 1893

The World's Columbian Exposition in Chicago proved to be one of the most significant cultural events in the Gilded Age. The massive fair encompassed 187 acres, visited by enormous crowds of spectators—ranging from farmers to academics—resulting in a dynamic impact echoing throughout American culture.[69] Joining the throngs were Native Americans and other indigenous people from around the world—some driven by the same popular appeal as everyone else. Others, however, were coaxed into attending by offers of economic gain from entrepreneurial endeavors. Commercial enterprises combined with fair organizers in attempts to bring indigenous peoples to the fair as living exhibits—an established tradition continuing in varying forms at dozens of fairs and events in ensuing decades. Starting in 1891, anthropologists and archaeologists began the task of gathering a massive amount of material for the planned displays in Chicago. As many as one hundred collectors across the globe worked to collect objects for display at the fair. The official anthropological exhibits at the exposition were largely crafted by Frederic Ward Putnam, then a curator at the Peabody Museum at Harvard University, and an anthropologist named Franz Boas.[70] Over the next decade, the two men worked to dramatically expand museum anthropology in the United States.

In the United States, even before a series of prominent world's fairs, Americans were already fascinated by racial difference and hierarchies. The exhibits at the World's Columbian Exposition, however, introduced an untold number of visitors to the emerging fields of physical anthropology and archaeology.[71] Exhibits at the fair subtly and overtly underscored existing ideas of racial difference with scientific undertones, pointing to the emerging studies of human skeletons as key evidence for racial classification theories. Displays at the world's fairs of the late nineteenth and early twentieth centuries included not just human remains but also living people. Displays regularly focused on indigenous people, generally casting them as savage and primitive natives, traveling to the fair from faraway exotic lands. Fairs drew large audiences by temporarily hiring and employing native peoples—often instructing them to reconstruct a traditional village and dress in their native style on the fairgrounds.[72] Accompanying these displays were occasional displays of human skeletal remains and mummies. In both life and death displays, audiences came to see race as the central lens for understanding humanity. Exhibitions of the anthropological body thus worked to reinforce race as a classifiable and seemingly static feature of humanity.

Diverse audiences coming to the fairgrounds afforded anthropologists the opportunity to take more anthropometric measurements—sizing up living people in addition to the dead.[73] Boas was hired to organize exhibits of the First Nations people of Western Canada as well as to collect anthropometric data from indigenous people visiting the fair. Boas had a position on the faculty at Clark University before arriving in Chicago, a city rapidly recovering from a large and devastating fire.[74] Of German-Jewish lineage, he trained as a physicist and geographer but became interested in anthropology while conducting fieldwork on Baffin Island. His early work resulted in a book published in 1888, *The Central Eskimo*.[75] From there, he became curious about (if not obsessed with) the cultures of the Pacific Northwest. At the World's Columbian Exposition, Boas collected bodily measurements of indigenous people from virtually every region in North America. His vast work on the Shoshone and Siouan tribes, based on these measurements, took years to publish. The data collected for the Sioux (Dakota) represented the measurements of 1,431 individuals, 186 of whom were children. Workers gathering the data at the fairgrounds noted skin color, hair color and type, lip shape, ear location, and the development of facial hair before measuring bone structure. Years later, in 1920, anthropologists looking back at the recorded information concluded that the data clearly represented the racial features of a single tribal group. Despite this, some argued that the geographical location of the Sioux, who lived across a wide swath of the Great Plains, allowed for the mixing of racial characteristics through interbreeding and tribal mixing.[76] The anthropologist Louis Robert Sullivan offered a possible explanation: "Among anthropologists who seek to explain the diversity of the American Indian physically by proposing two migrations, the one of a short, short-headed type and the other of a tall, long-headed type, the Sioux are usually pointed to as the results of intermixture of these two types."[77] Despite collecting a massive amount of data and nearly thirty years of study, anthropologists still struggled to articulate the meaning and form of racial characteristics in North America and around the world.[78]

By the time of the fair, Boas was already experienced in collecting human remains. During fieldwork in British Columbia only a few years earlier, he had personally collected the skulls of dozens of Northwest Coast Indians, offering them for sale at five dollars a skull and twenty dollars for each complete skeleton. Shipping them into the United States under falsified invoices, he attempted to sell them to the American Museum of Natural History, where he later worked as a curator. In 1887, Boas successfully sold several crania from the Pacific Northwest to the AMM, for prices ranging from two and a half to five dollars.⁷⁹ Boas lamented around the same time, "It is most unpleasant work to steal bones from a grave, but what is the use, someone has to do it." Despite his reservation, Boas argued vehemently in favor of the overall utility of collecting skeletal remains in order to understand and teach issues of race. When he finally viewed firsthand the World's Columbian Exposition exhibits he helped plan, he expressed dismay over the fact that the skulls in his collection were relegated to a small glass case in the corner of the fairgrounds. He bemoaned that the skulls were "likely to be overlooked by nine out of every ten visitors."⁸⁰

Popular anthropology displays complemented exhibits featuring recent discoveries from the cliff dwellings in the American Southwest. Audiences read about the mummified remains in newspapers, but they were also learning about ancient mysteries in the American West through an expanding literature. Ranging from romantic accounts of ancient history to informational guidebooks, these works helped develop a broad cultural awareness of fascinating discoveries taking place in the region. Audiences were drawn to the fair, at least in part, by accounts of ancient and mysterious places of the world—with the opportunity to view actual artifacts and genuine human remains from these places. Adolph F. Bandelier, previously appointed by the AAAS to explore the ruins of the American Southwest, published a best-selling, fictionalized account of the history of the region, *The Delight Makers*, in 1890. Though the book was not well received by critics, its popularity broadened awareness of the prehistory of the region.⁸¹ Other popular literature included *Some Strange Corners of Our Country*, a work of travel boosterism for the region by Charles F. Lummis and *The Land of the Cliff Dweller*, a popular history by F. H. Chapin also appearing in 1892.⁸² Taken together, these works fueled interest in both the contemporary and the ancient American Indians in the region—attracting audiences especially to see the American mummies whenever they were displayed at fairs before they arrived at museums in cities like Philadelphia, Chicago, and San Francisco.

The Wetherill brothers, who had continued to collect in the West, sold a collection of about one thousand "specimen" human skeletons, skulls, and mummies—filling exactly forty-two boxes—to C. D. Hazzard of the H. J. Smith Exploring Company, based in Minnesota. Hazzard moved the entire collection to Chicago and exhibited it at the fair as a replica cliff dwelling. Further artifacts were displayed in a replica "mountain," which, though being closer in resemblance to an outsized papier-mâché hill, attracted audiences that had not originally intended to tour the replica cliff dwellings to view parts of the collection. Following the fair, the collection was loaned to the University of Pennsylvania.[83] A few years later, donors purchased the collection, splitting it between the University of Pennsylvania and the University of California.[84]

Hazzard's collections included human remains representing, at minimum, eighty-six different individuals. The remains ranged from complete mummies to small chunks of preserved flesh and tissue discovered in caves. The collection also included bone fragments and bleached skeletons found throughout the arid climate of the American Southwest. Human remains represented slightly less than 10 percent of the collection Hazzard brought to Chicago, yet they were the most noteworthy part of the private collection.[85] Visitors to the displays encountered a series of mysterious bundles in which, hidden from view, were mummies of infants wrapped in cradleboards. The adult mummies were even more dramatic. Many with dark hair, dried skin, and exposed teeth, they would have undoubtedly captured attention. The mummification process left the bodies dried and appearing almost emaciated—leaving toenails, ribs, and hand bones exposed. Wrappings had blended with skin and dust. The mummified bodies were in varied states of preservation, many appearing gruesome and transformed by time into an almost unrecognizable state. Where possible, exhibitors pulled away the matting covering mummies to expose the remains to audiences.[86] Hazzard, who fell ill during the fair, decided to divest himself of the collection following the exposition, particularly as its immense popularity ensured his ability to find a buyer.[87] Frank Hamilton Cushing, a highly regarded anthropologist affiliated with the Bureau of American Ethnology, estimated the collection as one of the most important available to illustrate the ancient history of the American Southwest. Cushing noted, in particular, that the exceptional preservation of the collection, as well as its variety, could teach scholars about the spiritual, artistic, and everyday life of the ancient Pueblos.[88] Adding to the value of the collection, of course, was the series of naturally mummified bodies, many still wrapped in original burial materials.

Adding to a mock cliff dwelling built entirely with the Hazzard Collection in mind, fair workers fabricated the "largest artificial mountain ever constructed" to display the material collected by the Wetherill brothers for the state of Colorado.[89] Visitors walked through a large entrance and were faced with a full-scale representation of the "craggy vastnesses of which many of the finest cliff dwellings" were found. Culminating with a walk through a reconstructed canyon, visitors entered the most provocative portion of the exhibition, which featured "some thousands of examples of the weapons, cooking utensils, implements and mummified remains of this prehistoric people."[90]

One volume documenting exhibits on the fairgrounds described in some detail the displays organized by private corporations or individual entrepreneurs, most of them hoping to sell "personal collections" at the end of the exposition. A caption underneath a photograph featuring men and women walking into the reconstructed mountain reads, "The visitor was introduced to a large exhibit of the mummified remains and domestic relics of the Cliff-Dwellings, the oldest semi-civilization of the Western Continent." It continued by describing the exhibition as "so skillfully arranged that the visitor to the display seemed to be standing in the very midst of the real ruins, and shaking hands, as it were, with the dusty remains of a people who played their part in the drama of the world more than a thousand years ago."[91] Human remains were becoming increasingly central in emerging visions of the prehistoric past of North America. Without fail, however, race was used as the central qualifier in communicating the significance of these bodies—their stories of a distant and exotic past essentially made for an intriguing setting for racial storylines.

The display of cliff dwellings at the 1893 World's Columbian Exposition was so popular that it was replicated, on an even larger scale, at subsequent international expositions. Before reappearing at major international expositions, however, the cliff dwellers were placed on display in a new exhibit of the Hazzard collection at the University Museum of the University of Pennsylvania. Ten mummies were the central attraction: "Naturally," one local newspaper read, "the most interesting portions of the collection to the average visitor are the exhumed bodies of the wonderful people themselves."[92] Another newspaper imagined that the mummies were, in fact, "hold[ing] a reception" at the opening of the exhibition.[93] Despite having been advised that the mummies were "not pretty things to look at," large audiences poured through the galleries, many of them no doubt already familiar with the collection that had been displayed so prominently in Chicago.[94] The media in Philadelphia was immediately captivated by the exhibition, promoting the relics of the cliff dwellers as evidence of the "contributions of the history of the American race, and the story of a new Egypt—a new Babylonia." Newspapers continued by reminding readers that the story of the cliff dwellers "was unfolded here in America, to take its place beside and confirm the Peruvian record of the early life of man on this continent [sic]."[95] They suggested, "Many prolonged visits will be required in the Museum to enable one to become even moderately familiar with all the manifestations of primitive life and industry displayed in the collection."96 By the 1904 fair, the exhibits of cliff dwellings, placed alongside living individuals from the Hopi and Zuni tribes, were so abundant that the gallery was described as capable of "form[ing] a complete Exposition in itself."97

By 1904, massive audiences in Chicago, Philadelphia, and St. Louis had viewed mummified remains discovered in the American Southwest. Visitors at museums and fairs were presented with mysterious and grandiose narratives about ancient history with the presentations of American mummies, but the lingering issue of race and racial classification surrounded these stories at every turn. Members of the public were invited to view the remains and imagine the past, contemplating their own connection to the bodies. Scholars and scientists, too, recognized the significance of these newly acquired collections, especially the American mummies brought together initially for display at the world's fairs. The permanent transfer to museums contributed to the gradual growth of museum collections of remains. Exhibits displayed at fairs held in the United States between 1893 and 1904 alone resulted in human remains being sent to museums in San Francisco, Philadelphia, Chicago, Denver, and Washington, DC. Competing collections appeared to be changing just as rapidly as the ideas that surrounded them.

FROM THE ARMY MEDICAL MUSEUM TO THE SMITHSONIAN

As AMM collections moved to the Smithsonian Institution in 1897, Smithsonian administrators proposed an internal, institutional reorganization into three separate departments: anthropology, biology, and geology. Each department was under the supervision of a head curator. Curators sought out new collections that represented the established or emerging fields of each respective discipline, while also seeking to hire new curators to serve as caretakers for existing collections. The anthropology department, in particular, rapidly attempted to reassess existing collections into an intellectual framework that could guide new collections. The museum's annual report described the anthropology department:

> There are a number of sections that have not yet been assigned to any division, remaining for the present under the direct supervision of the head curator. Moreover, the classification of material and the division of work among the various members of the present staff, so far as it has progressed, is largely tentative, owing to the staff being composed of specialists in limited portions of the field of anthropology; this necessitates a somewhat arbitrary classification and organization. As the various branches of the work develop, and increase is made in the number of curators, reclassification of material and readjustment of the force will gradually lead to satisfactory and permanent organization.[98]

As the Smithsonian defined departments and acquired collections, it intended to shape and develop academic disciplines by acquiring the raw materials to buttress future research. This goal neatly complemented the idea that certain collections, including human remains, were both a limited and a valuable resource. Even if the exact intellectual meanings for these materials were unclear, collections like skeletal remains needed to be urgently acquired now in order to unlock key ideas about race and history, ideas deemed central for future scientists. The anthropology department at the Smithsonian, in particular, had managed to bring together a large collection from a wide range of sources around the globe. In 1898, the department proudly declared that it had acquired 1,441 separate accessions, containing upward of 450,000 individual specimens.[99] In other words, in that single year, the Smithsonian's anthropology collections acquired enough material to establish what might have made another respectable museum. The massive growth in the collection is still more impressive given that the museum pointed to a "meagerness of funds" as having handicapped its efforts in obtaining specimens. Further, the Smithsonian Institution as a whole experienced a dip in attendance after the outbreak of the Spanish-American War.[100]

Contributing to the astonishing expansion in the collections during the year 1898 was the acquisition of human remains from the AMM. Ultimately the Smithsonian received "a collection of 2,206 human crania, representing mainly the Indian tribes, ancient and modern, of North America."[101] At the time, physical anthropology was still an ill-defined field, tied together by a large and diverse body of scholars and quasi-intellectuals interested in race theory and a smattering of scholars and amateurs interested in studying the ancient cultures of the world. Nevertheless, the moment proved to be a turning point, prompting the nation's largest museum to determine its level of commitment to physical anthropology as a discipline.

With new collections now in hand, Smithsonian leaders felt emboldened to make a major decision. In 1903, Aleš Hrdlička, a Czech-born anthropologist, became the first curator of physical anthropology hired by the museum. One of seven children, Hrdlička and his family migrated to New York when he was thirteen years old. His scientific training began in earnest when he enrolled at Eclectic Medical College, before continuing at Homeopathic Medical College in New York. In 1896, he traveled to Paris, where he studied with prominent anthropologists and physicians who further influenced his thinking. A stubborn yet sharp intellectual force, he rose to new heights at the national museum—just as it acquired a major collection of skulls and bones from around the world. After arriving in Washington, DC, Hrdlička became a critical figure in the formation of physical anthropology in the United States, but his legacy is seriously fraught. He was intense, incendiary, and extraordinarily prolific. The Smithsonian, under his leadership, became a central core around which other human remains collections gravitated. Despite his intellectual influence, Hrdlička's behavior was often interpreted as cold and sometimes even combative. His mere presence fueled competitive sentiments and tension connected to ultimate control over bone collections. A voracious writer and defender of his particular approach to the science, he not only had access to vast collections in the United States but was frequently granted first access to view the most important skeletons and fossils in museum collections abroad.

In Hrdlička's leadership capacity, the curator founded the *American Journal of Physical Anthropology*, and given his connection to both the journal and the Smithsonian, he commanded much of the discipline of physical anthropology and the study of race and history through the study of human remains. His drive to collect and organize skeletons was based on his theories about race; Hrdlička adopted the racial classification scheme of the nineteenth-century French naturalist Georges Cuvier. Following Cuvier, Hrdlička extended the idea that humankind might be divided into three basic stems—white, black, and yellow brown. Beyond that simple division, the racial tree of humanity was confusing and unclear. Responding to the desire to understand and classify particular "subracial" units, museums collected and classified—hoping that repeated measurements would reveal patterns in humankind's confounding racial tree.[102] Alongside Earnest Albert Hooton at the Peabody Museum of Harvard University, Hrdlička was the most important physical anthropologist in the first half of the twentieth century. Despite the many differences between the two men, they came to exemplify the emergent field of physical anthropology as manifested through the careful study of comparative human anatomy.[103] The two men, and Hrdlička in particular, were especially concerned with the concept of "ancestry," which was understood at the time to mean both the deeper implications of human evolution and also the classification of humanity in racial terms.[104] Hrdlička proved the more challenging of the two—an unabashed iconoclast, he nevertheless clung to Procrustean solutions to scientific problems—his stubbornness expressing itself through both personality and intellect.

Hrdlička came into his position with the benefit of a newly acquired collection from the AMM, including specimens from around the world and already the envy of many young anthropology departments. He and the Smithsonian both benefited from large and diverse bone rooms filled with skeletons. Hrdlička gained experience working in the Department of Anthropology at the American Museum of Natural History before becoming a curator at the Smithsonian. Hrdlička, following his European, French-centered training, spent his career downplaying the importance of genetics and emphasizing morphological characteristics of human beings. Although he had proved central to the overall growth of the field—and was the most important figure in shaping the practice of collecting skeletal remains for museums in the United States—toward the end of his career, the field seemed to pass him by. Hrdlička had high demands for both his employees and other scholars, and his frequent challenges to their attitudes and ideas were sometimes viewed as arrogance. For a time, he was a member of the American Eugenics Society and worked with a number of the most significant eugenicists in the United States, tying the field together with the newly materializing discipline of physical anthropology. For all his faults and commitments to shortsighted ideas, Hrdlička had a knack for collecting, measuring, observing, and organizing human bones. Overall, he generally encouraged the growth of American physical anthropology and morphological studies of both living humans and collections of human remains housed in museums.[105] His legacy is not beyond reproach, however, and his complex views toward race and evolution have since been roundly criticized. In a private letter written during the midst of

Hrdlička's career, one cultural anthropologist described him as "never [having] produced scientific work above the level of mediocrity."[106] Others, certainly including many in the emerging discipline of physical anthropology, disagreed. Regardless of personal opinions of Hrdlička, however, many scientists and amateurs proved willing to submit boxes of skeletons to the Smithsonian.

Upon arrival at the Smithsonian, Hrdlička quickly began a project, on a global and multifaceted scale, to collect and study human remains, using almost any means necessary. Despite this apparent growth, few scholars in the United States were yet trained in the young field of physical anthropology. Most scholars of this era focusing on race and physical anthropology trained in either medicine or biology. Although Hrdlička did much to advance the field of physical anthropology, he maintained that the best training for physical anthropologists was in medicine, even as the two fields drifted apart over the course of the first half of the twentieth century. In Europe, Hrdlička's influence was compared to one of his mentors, the French anthropologist Paul Broca (1824–1880), who developed centers for physical anthropology in Paris.[107] Hrdlička, having spent time training with Broca in France, hoped to emulate such institutions in the United States, with the Smithsonian forming the most important site for teaching and research in the field.[108] Beyond institution building—if the races of humankind were to be understood on the axis of white, black, and yellow brown—the United States in the late nineteenth century seemed to be the ideal testing ground to prove and better comprehend such an assumption. Bone rooms could put new racial theories to the test.

RACIAL THEORY AT THE DAWN OF THE TWENTIETH CENTURY

As new evolutionary theories came to dislodge and replace older ideas about polygenesis and monogenesis, racial classification theories became more complex. Although Darwinian evolution had come to dominate the natural sciences by the turn of the century, anthropologists like Hrdlička and his British contemporary Sir Arthur Keith viewed Darwinian natural selection as problematic.[109] French anthropologists who guided much of the thinking in physical anthropology over the previous fifty years had largely aligned their views with the French naturalist Jean-Baptiste Lamarck (1744–1829), devising a theory of evolution that, for a time, competed with Darwin's for preeminence in continental Europe. Lamarckian frameworks for human evolution largely dominated French anthropology in the second half of the nineteenth century.[110] The early scholarly reception of Darwinian theory in the United States proved less contentious than in France, and many scientists in the United States worked to advance a more secular evolutionary theory. Despite the general movement away from the polygenesis/monogenesis dichotomy, the question of what ideas would buttress racial classification theories remained unanswered. In the decades following the Civil War, an untold number of wide-ranging racial classification schemes were published in the United States; some of these schemes were supported by research conducted on collections of human remains, while many were based purely on speculation. Many scholars collecting, measuring, and observing human remains avoided these debates entirely, content to create charts and graphs without fully extrapolating their data into narrative theories about race.

In the United States, debates about race soon centered on the discourse surrounding the effect of the environment on the human body. To what extent, anthropologists came to wonder, did the wilderness of North America shape American Indians since their arrival on the continent? Scholars also debated the long-term implication of European and African presence in the United States. Through the measurement and collection of skulls, some argued for the stability of races. Those with positivist leanings argued that these stable races could be arranged within consistent taxonomies of races. Others came to the position that while representative examples of races may have existed at one time, the gradual intermixture of cultures and populations would continue to confuse the stability of races.

Two seminal figures in the field of anthropology, Hrdlička and Boas, agreed that the environment had a direct impact on the human body. As evidence, both scholars directly explored the human body, particularly considering how the bodies of living immigrants had changed over several generations and, significantly, how the bodies of American Indians may have changed since their arrival in the New World. The two arrived at different conclusions that shaped much of the discourse in anthropology over the next generation. For a time in the early twentieth century, both figures worked to build collections of human remains through opportunistic acquisitions of skeletons for museum collections and systematic measurements of living individuals.

In 1899, Boas published an article in *American Anthropologist* titled simply "Recent Criticisms of Physical Anthropology." Boas had, by this point, positioned himself as a key leader in the field of American anthropology. As noted earlier, he had also actively collected human remains, once even going so far as to use a photographer to distract local natives while he pillaged graves.[111] Despite this background, Boas was hardly just another profiteering grave robber; he possessed a serious intellect and was genuinely interested in the problems of race and culture in a wide variety of contexts. Boas was a critical player in the rise of museum anthropology, but his relationship with the museum as an institution had become somewhat strained, leading to his eventual departure from the American Museum of Natural History for Columbia University.

In the *American Anthropologist* article, Boas actively defended his research against the criticisms, already apparent before the turn of the century, that the utility of physical anthropology was only minimal and that the process of gathering information about body measurements was fundamentally flawed by an overall lack of clarity in racial theory. Boas defended the field of physical anthropology overall; his article begins with an explanation as to why anthropologists were interested in skeletal materials:

> One of the principal reasons that led to a more detailed study of the skeleton and to a tendency to lay the greatest stress upon characteristics of the skeleton, was the ease with which material of this kind could be obtained. Visitors to distant countries are likely to bring home skeletons and parts of skeletons, while not much opportunity is given for a thorough examination of a considerable number of individuals of foreign races. The difficulty of obtaining material relating to the anatomy of the soft parts of the body has had the effect that this portion of the description of the anatomy of man has received very slight attention. In comparatively few cases have we had opportunity to make a thorough study of the characteristics of the soft parts of the body of individuals belonging to foreign races.[112]

Skeletal material, in other words, was in large part collected out of convenience and assumed stability in form. Boas continues his article by arguing that skeletal evidence allowed anthropologists the opportunity to understand not just living races but also long-extinct races and populations. Although Boas acknowledges the desire within the discipline to study prehistory through human remains, he primarily emphasizes the need to utilize physical anthropology to understand race; his article underscores the aspiration to understand racial groups rather than individual specimens. This is important, he argues, because though the discipline of physical anthropology had been critiqued for the inconsistency of individual measurements, the overall consistency of what were argued to be clear groupings points to the existence of a number of races of humankind. For Boas, like other scholars of the period, the study of prehistory was less important on its own accord than it was vis-à-vis the desire to understand the existent races of humanity. Boas, quite unlike others in his field, showed a capacity for intellectual change as new evidence became clear. His position on this and other issues changed subtly in later years.

Where Boas began to break from many of his contemporaries was in his resistance to *strict* schemes of racial classification. Boas believed that while understanding racial relationships might prove to be possible, strict racial classification was only an imagined reality. Individual bodies, he noted, might be shaped by their surrounding environment and through various cultural practices. Boas argued, "Each social unit consists of a series of individuals whose bodily form depends on their ancestry and their environment."[113] Boas's environmental determinism reflected something of his overall theoretical framework for the development of human culture as well as the human body.

Although Boas believed that heredity exerted greater influence than environment in the shaping of the human body, he argued that the evidence was incomplete. More measurements would need to be taken and more remains would need to be collected before an answer to the question could be fully articulated. "The statistical study of types will," Boas noted, "lead to an understanding of the blood relationships between types."[114] He concluded that only by gathering more information about the diversity of the human form, together with the complete study of cultural habits and language, would keys emerge to unlock the secrets of humanity.

By the early 1910s, Boas's studies exploring the effects of inheritance and environment on immigrant populations pushed him to intensify his attacks on racial classification. The course of his ideas eventually influenced the entire field of anthropology. At this early stage, however, his uncertainty with the meanings of early measurements and collections simply pushed the field to gather more data and more bones. Around the turn of the twentieth century, Boas reflected on problems inherent in existing theories on race and culture, fundamentally undermining a myriad of poorly defined yet dominant theories that had inspired the early collecting of human remains.

While both Boas and Hrdlička were advocates for environmental determinism, the two disagreed on how quickly the environment influenced the human body. Hrdlička proved more willing to put races into strict typological schemes even as Boas's vision of race became more fluid. As Boas increasingly turned his attention to questions of culture and anthropological theory, Hrdlička turned his own gaze deeper into human history.[115] Whereas Boas departed the museum setting for the university, Hrdlička stayed at the Smithsonian for several more decades, his tenure ending at the museum when he died. Boas's emphasis within anthropology shifted intellectually away from questions regarding the human form, but Hrdlička's obsession with the quest to draw ideas from the bones of the dead only seemed to deepen. Driven by questions of science, unmitigated ego, and widespread fascination with race, scientists ensured that bone rooms would grow at an unparalleled rate.

By 1900, museum professionals developed various systems for organizing collections of human remains. Theories about race, gender, anthropology, intelligence, and medicine all directly influenced the practical problems faced by museums in storing and organizing their human remains collections. Debates connected to these questions, and the degree to which certain scholars emphasized some qualifiers over others, occasionally resulted in heated discussions. Seemingly inconsequential details—including skull measurements or whether certain catalogues of collections should identify the known medical history of the deceased—were considered critical to some scholars while being simply irrelevant to others. Generally, human remains collections were conceived as providing research opportunities in areas such as comparative anatomy, anthropometry, morphology, pathology, kinetics, and taxonomy, among others. Complicating this process were the ultimate goals of such varied research programs; scholars aimed to answer questions about race, medicine, physics, chemistry, biology, evolution and development, anthropology, and, increasingly, human prehistory. With such an assortment of questions came the need for more and more data to make sense of the implications drawn from human remains collections. Although many competing theories about race and evolution that were taking hold in the middle of the nineteenth century were short-lived, they inspired an all-out flurry of collecting that influenced all branches of anthropology in the United States.

Despite the influence of multiple waves of racial classification theories, human remains were largely organized and collected around the basic tenets of racial group and sex. In 1900, Hrdlička published an article in the *American Naturalist* explaining his own system for organizing remains:

> For anthropological and zoölogical collections of bones, probably the best general rule is to keep, in appropriate series, all the bones of each skeleton together, minus the skull. Each bone should bear the number of the skeleton. The skulls of the same tribe of people or species of animal are kept together, heading the series. Each distinct group of skulls and skeletons in a collection is divided and arranged in at least three groups: the children or young, and grown individuals, separated into males and females. In large series the embryos, adolescents, and very old may be advantageously separated from the others.[116]

Hrdlička continued, "The anthropological collection as a whole is arranged on the basis of race and type, and further subdivided according to geographical distribution."[117] Individual variation within racial groups, including such factors as gender and age, were subverted by the emphasis on racial type.

With the growing desire to obtain human remains came obvious obstacles to procuring such commodities. At the turn of the century, these efforts often went hand in hand with burgeoning strategies for museum exhibition. As Hrdlička was busy organizing and numbering his skeletons, the American Museum of Natural History (AMNH) in New York was completing a series of new exhibits intended to "illustrate the different types of the human race."[118] While publications like Hrdlička's manual on collecting and organizing human skeletons did little to raise the overall awareness about the project to collect and interpret bones in the museum, exhibitions like those planned for the AMNH in New York and world's fairs elsewhere helped broadcast the nature of the bone-collecting project to a diverse audience.

COLLECTING THE LIVING AND THE DEAD AT THE ST. LOUIS WORLD'S FAIR, 1904

Efforts to display *living* humans at museums and fairs in the late nineteenth century proved successful for the organizers of such exhibitions. While these types of exhibitions were not cheap—living indigenous peoples needed to be transported, housed, cared for medically, and fed—they attracted a significant amount of attention from fairgoers and media alike. Although earlier efforts to display the living had taken place in the United States, the display and exploitation of living indigenous people was most prominent at the 1904 St. Louis World's Fair. Intended to celebrate the hundredth anniversary of the Louisiana Purchase, the Louisiana Purchase Exposition of 1904 continued the tradition of popular displays of ethnographic material culture and live human exhibits intended for large audiences. The fair, which encompassed over 1,240 acres, built on earlier ideas for "living exhibits" dating back to the 1876 Centennial Exposition in Philadelphia. In 1876, Congress had balked at the proposed price tag for an exhibition bringing together American Indian individuals from around North America, but private commercial endeavors routinely had stepped in, profiting by coaxing native people to serve as living exhibits. Fairs held in Europe demonstrated both enormous success and deep interest in the living displays of natives, providing an example for later fairs in the United States to follow.[119]

While some commercial endeavors proved successful, attempts to gather representatives of American Indian cultures for fairs in the United States produced only mixed profits. Billed as the final opportunity to view dying races firsthand, the 1898 Trans-Mississippi Exposition was promoted as the "Last Great Congress of the Red Man." What was envisioned as a massive gathering of Native Americans from around the continent resulted in only a modest gathering of four hundred individuals from the Plains and Southwest. Fair organizers attempted a similar gathering in 1901 at an exposition in Buffalo.[120] By 1904, therefore, the concept of displaying exotic bodies had become firmly entrenched in the minds of fair organizers, even if past efforts resulted in dubious—and occasionally disastrous—outcomes. Those who organized the St. Louis fair, however, were more successful in the planning and execution of profitable exhibits, and the exposition featured displays of indigenous people from around the world, or "living fossils," as examples of humankind's evolutionary past.[121]

For the indigenous peoples, travel to new places meant in some cases sickness and death. In 1904, Hrdlička, the Smithsonian's new curator of physical anthropology, was determined to take advantage of such unfortunate inevitabilities. Whereas anthropologists working at the 1893 exposition in Chicago collected mainly measurements of the living, Hrdlička believed the 1904 fair might result in new opportunities to actually collect those who perished. Hrdlička had proven to be an opportunist in the past, once working with Boas to collect and study the remains of six Eskimos brought to New York City for what was intended to be a temporary display of the mysterious individuals of the world's northernmost culture.[122] Hrdlička himself had helped orchestrate some of the Smithsonian displays for the St. Louis fair, and he decided it would be wise to make a prolonged visit to the fairgrounds. Upon his arrival, he set about collecting the brains and skeletons of indigenous people who had died, largely from communicable diseases. Hrdlička performed numerous autopsies while visiting the fair and managed to collect over two hundred specimens. He sent them all to the Smithsonian. He also made numerous facial casts and photographs of the individuals in the living exhibitions of the fair.[123] The nature of their display at the fair, coupled with their treatment after death underscores the degree to which indigenous bodies were considered to be commodities by the scientific community in this era. This kind of opportunistic approach to bone collecting, the creation of organizational frameworks, and support from museum administrators allowed the Smithsonian's bone collections to grow to unparalleled size.

. . .

While human remains arrived at museums through professionalizing channels in archaeology and physical anthropology, museums also began acquiring skeletons from other diverse and enigmatic sources. Medical officers, amateur collectors, looters, pothunters, and treasure hunters worked over many of the same sites later studied by professional archaeologists and anthropologists. Although diverse groups, they were ultimately connected by the fact that many of their finds would be deposited in shared collections at permanent museums. In fact, this transition into permanent museums was occurring only as the field of physical anthropology—still centered on racial classification theories—was starting to emerge as a legitimate scientific discipline in the United States. Despite lacking a totally systematic strategy for collecting human remains in either medical or natural history museums, the project to scientifically organize and classify races through bones became ensconced in several major institutions. Due at least in part to the display of human remains at the turn of the century, Americans were becoming familiar with race as a scientific concept. Major exhibits at fairs and museum, works of literature, and media attention publicized museums' efforts to collect skeletons and encouraged amateur collectors to gather skulls. Unlike when collecting or looting art or rare burial goods, bone collectors never found a viable market for skeletons that had been taken from the ground. Although fantasies about striking it rich rarely became reality, many skeletons and mummies collected by relic hunters did eventually make their way to museums. These remains frequently arrived with limited provenance information, confounding contemporary efforts to study or repatriate these collections. Remains collected by professionals like Boas and Hrdlička complemented the privately collected bodies and body parts. Despite previous collaborations, the two were gradually arriving at competing ideas about bone collections and race. Once friends and collaborators, Hrdlička and Boas increasingly grew apart and became increasingly critical of each other's competing theories. Not only theoretical in nature, the discomfort in their relationship extended into a tense rivalry over controlling bone collections themselves.[124]

Following the extremely successful displays of human remains on the fairgrounds of international expositions, museums developed their own exhibitions featuring mummified remains from the American Southwest as a prime attraction. Private and public enterprises brought exhibitions of human remains together to popularize a tapestry of scientific and pseudoscientific ideas about race and, to a lesser extent, narratives about human history. Commercial bone exhibits at international expositions proved short-lived, and many bodies were ultimately sold or donated to museums. Meanwhile, as the mummified remains of the cliff dwellers captured popular attention, other human remains that were thought to reflect existing races of humankind were quietly arriving at museums from the American West and around the globe.

When Hrdlička arrived at the Smithsonian in 1903, he noted that he would need the help of several types of people to build the collections. In an internal memorandum, he first lists physicians in Washington, DC, specifically those "who have charge of the morgue," as his most important connections. He was also quick to note the role of the army, the navy, and Indian service agents in the acquisition of existing national collections of remains. Similarly, Hrdlička cited the role of "Missionaries and Consuls," exchanges with other museums, and his own work as an archaeologist. He fully expected this kind of regular but piecemeal pipeline of skeleton collecting to be exploited to its fullest potential.[125] The plan seemed to be working.

In 1904, Hrdlička published a brief guide for collecting and preserving human remains. The guide was written with museum professionals in mind, but it was also intended to encourage the development of private collections, which might later result in museum donations.[126] Hrdlička's guide was in the tradition of older articles appearing in magazines for scientific elites throughout Europe.[127] It detailed practices connected to collecting and preserving human remains while, importantly, reinforcing Hrdlička's vision for the collection. The guide starts with clear instructions for dividing human remains collections into three general categories: "the whites and other civilized peoples," "those among primitive peoples," and "those of extinct peoples and early man."[128] Over the ensuing decades, Hrdlička witnessed the rise and fall in significance of the first two categories while observing and contributing to the continual growth of the study of ancient humans. In the early years of the twentieth century, the division between whites and other races that were thought to be more primitive defined and dominated the field of physical anthropology. Hrdlička's personal obsession with collecting and organizing the remains of races from around the globe meant that the Smithsonian was eager to accept donations from a wide and diverse range of colonial collectors around the world.

The AMM story and the Smithsonian's reorganization indicate that many prominent figures in medicine and anthropology in the United States were attempting to define the newly acquired bodies on museum shelves. With the arrival of Aleš Hrdlička at the Smithsonian, the frustratingly elusive questions of race and history were given new life in the ensuing decades. Media attention, successful exhibitions, and the coming of age of a new generation of scientists set the stage for a growing movement to collect bones.

CHAPTER 2

SALVAGING RACE and REMAINS

Hard and steady rain fell in northern Minnesota during September 1918. Frances Densmore, fifty-one years old, professionally dressed, and serious-minded, set herself to work with the Chippewa Indians. Densmore spent most days diligently recording songs, notwithstanding the dreary weather that month. As her work with the Indians was nearing an end, receding water above a dam washed away the riverbank, exposing long-forgotten relics. Densmore knew little about archaeology, but she believed the findings "seemed too interesting an opportunity to slight." Besides, she remarked, "The material was being picked up rapidly by those who would never make any use of it."[1] Human bones were strewn among burial objects, freshly exposed by the recent heavy rains. This was Densmore's first experience collecting skeletons, but the idea that bones and burial goods needed *salvaging* for those who would *make use* of them caused her to react quickly. The bones were the only remnants of a living human being, but they had also become a commodity. As the driving rain continued to fall, Densmore implored a local man to help gather the remains in haste. Later, a young boy gave her several additional skull fragments found at the same place.

Although she believed the bones found might be valuable, she considered the skull to be the most significant. Even an ethnomusicologist was aware of the prevalent idea that skulls possessed the most important clues about the deceased.[2]

Densmore soon shipped the bones, along with the pottery fragments she found nearby, to the Smithsonian.[3] It is unclear how the local Chippewa might have reacted to the natural exposure of bones and burial goods, but Densmore's letters lay bare her understanding of the bones of Native Americans as tools for science, a frame seemingly unknown to others who may have collected bones for some other purposes. Like Densmore, a wide array of anthropologists and amateurs reacted similarly when finding indigenous human bones or mummified bodies, shipping them to major museums or one of the many smaller regional or university museums. Numerous amateur collectors possessed only vague knowledge as to why exactly museums would collect human remains; they simply collected them and sent them to museums. Densmore, like a surprising number of anthropologists not typically remembered for collecting bones, maintained only a marginally clearer understanding of human remains' use to science. Nevertheless, the perceived value of these artifacts drove enough continued action to result in their ongoing collection. The pattern of occasional, accidental, and unorganized collecting pushed museum collections to greatly expand over time. The trend in collecting human remains grew beyond professionals who sent discoveries to museums for preservation, study, and display. Indeed, extensive looting from important archaeological sites, including those with important skeletons and mummies found throughout the American West, pushed the federal government to take action to prevent further damage to historical sites and landmarks. Without legal or professional guidelines, allowing haphazard bone collecting to persist created ethical problems with which museums still wrestle today.

Weeks after Densmore's initial discovery in the early twentieth century, she wrote to the museum "with some anxiety," curious about the remains.[4] She wondered if the bones arrived at the museum safely but was also inquiring if they were of any value to science, hoping they might document and preserve a secret of race or the past. Densmore believed important pieces of knowledge were at stake at that moment in time; evidence about race and human history may have washed away had she not acted to save the rain-soaked bones. After some delay, Aleš Hrdlička, curator of physical anthropology at the Smithsonian, responded. Hrdlička, buried deep in his laboratory within the museum, appreciated the anthropologists and amateurs who sent packages of skeletons to the Smithsonian on occasion, though he was also often wary of the claims of significance and stories ascribed to the bones of Native American "chiefs" and "warriors" frequently turning up in his office. Accidental discoveries of human remains were common, especially as farmers tilled new fields and explorers mapped new territory throughout the American West. The rapidly increasing number of boxes sent from scholars, scientists, and self-proclaimed colonial adventurers meant that Hrdlička could respond only briefly to each sender. These remains, he remarked to Densmore, "while too fragmentary to be of any anthropological importance, ... show a number of artifacts which will well justify their preservation."[5] A sharp object punctured the bones, in some places even exposing the marrow, Hrdlička explained to Densmore. Hrdlička believed the American Indians who buried them, hoping to release any evil power embedded inside, intentionally shattered the bones.[6] The Smithsonian carefully wrote catalogue numbers on each bone and filed them away deep within the museum. Although Hrdlička seemed to have no immediate plans to study the bones, another scholar, it was presumed, might find the remains to be scientifically significant at some future, unknown date.

While Densmore was not an archaeologist, she believed it her responsibility both to collect the remains *and* to submit them to the Smithsonian.[7] Generations of like-minded Euro-Americans assumed Native Americans to be rapidly disappearing, and the main tenet of "salvage anthropology" was to collect and record American Indian culture and race. The scholars aspired to "salvage" whatever might be preserved, from pottery to linguistic data to human bones. Speaking to the importance of building a collection of human remains, Hrdlička wrote to his supervisor, "If it is urgent to gather data on the language, religion, and customs of people who are disappearing, it is surely quite as urgent to secure a physical record of the same groups, records which will always remain the most substantial criterion of their classification."[8] If the races were vanishing, it was argued, so too was the racial record held in their bones.

Not all anthropologists of this era considered themselves to be salvage anthropologists, but the manner in which many in the anthropological community collected and gathered materials in this era was deeply influenced by the desire to preserve all they could of peoples they believed were disappearing, rather than constantly changing and adapting. The campaign to preserve and collect was viewed as a *race against time*; bone empires benefited from this powerful sentiment by conceptualizing indigenous and ancient bodies as a limited and scientifically valuable resource. For some anthropologists, this framework encouraged the idea that rare discoveries of human remains—whether they be found strewn across a riverbank in Minnesota or deep inside an Australian cave—were potentially valuable additions to growing collections. It might be argued that salvage anthropology was less a theoretical approach to the discipline than an absence of theory. Indeed, the concept of salvaging a dying set of American Indian cultures stretched so far beyond the small community of anthropologists that it is difficult to treat the efforts as strictly an anthropological practice or even intellectual framework alone. Instead, the movement to "salvage" cultural and racial data through preservation of art, material culture, linguistic information, physical measurements, and even skeletons extended beyond intellectual circles. The broad range of amateur collectors who submitted material to museums in this period suggests an extensive, yet vague, understanding of the museum's desire to collect certain specimens of vanishing peoples.

This was not just a popular notion, however, as scholars working around the world were influenced by this idea of conceptualizing primitive peoples around the world as rapidly succumbing to the onslaught of Western modernity. Just as scientists were becoming more interested in preserving the culture and bodies of supposedly disappearing races, the public was eager to embrace displays of their material culture and skeletons. The notion of vanishing "races" of modern man helped distill the less dominant notion of ancient, extinct races of humans—mysterious groups leaving behind architectural (read: archaeological) mysteries as well as confounding mummies and skeletons. The result was not only an increasingly large set of collections to fill the bone rooms of museums in the United States; the movement also pressed for a series of new ideas and legislative developments governing the acquisition and display of the human body.

The American Antiquities Act, approved by Congress in 1906, enabled critical new turns. The law aimed at penalizing amateur ransacking of archaeological sites in favor of working with professional archaeologists to preserve historical sites deemed of national significance. But in the face of the new law, during the first half of the twentieth century, skeletons continued to be routinely shipped to major museums in the United States. Bones were as often gathered in a manner resembling Densmore's experiences as trained archaeologists collecting them.

Shipments came from around the globe, and most were readily accepted, causing the size of collections to balloon. The moment proved to be the apex of museum collection building in the United States, when judged simply by the number of catalogue entries added to museum records. Despite the seemingly critical turning point manifest in the 1906 Antiquities Act, looters, amateur collectors, tourists, and tenuously connected professionals continued to freely collect, study, and display human remains well after the law's passage. The law created, at best, a weak and permeable shield around remains situated on federally owned lands. Nevertheless, it created the first federal guidelines governing collecting human remains and other antiquities in key parts of the United States and its territories. As such, it came to provide a legal framework that eventually regulated the acquisition of thousands of human remains discovered in large sections of North America. The law began to establish a precedent for the legal acquisition of archaeological material—including human remains—and it worked to discourage looting through fines.

Following the passage of the American Antiquities Act, scientists working with human remains also began an organic and gradual intellectual shift. Before the 1910s, most scientists working with human remains were interested mainly in racial science. Simply stated, race so completely dominated popular and scholarly discourse surrounding remains that even discoveries of ancient and even fossilized remains were framed by racial description. Whenever an unusual body was found, curious minds wondered how it might be understood in racial terms—as concepts, *ancestry* and *racial history* became intertwined. While some scientists and members of the public maintained an interest in human origins before that decade, it was not until a series of major discoveries around the globe occurred in the 1920s and 1930s that the interest in human evolution and prehistory expanded. During the early part of the twentieth century, the study of human evolution centered largely on the evolution of the races, with much of the work clearly demonstrating a racist underpinning veiled in the language of scientific authority.

These seemingly divergent themes—collecting and displaying human remains by amateur archaeologists and the course of the scientific shift from racial classification theories to human history—were intimately connected by the actual practice of collecting, studying, and displaying human bodies. This period provided a series of key turning points that ultimately changed the way human remains held in museums were collected, organized, and interpreted. As the American Antiquities Act gradually influenced the way human remains were collected in the field, scientists used remains to test existing theories on the subject of race and also began using bodies more intensely as tools to explore ideas about human history. At the same time, scholars internally debated the place of eugenics and scientific racism in the study of their collections. Before they could truly begin hashing out these later debates, however, scholars were making key decisions on how early human remains collections would be organized in growing museums—mostly in cities and university towns along the East Coast. The newly emergent leader in the field, the Smithsonian Institution, was among the first to publicly try to tackle the problem of organizing human bones into clear categories for science on a large scale. At the start of the twentieth century, the act of organizing bones on museum shelves represented something of a physical articulation of attempts to create a science to classifying humankind.

A STATE OF DISARRAY

In 1903, as scholars and government officials were busy lobbying for the Antiquities Act before Congress, Aleš Hrdlička first arrived at the Bureau of American Ethnology. Not only was he given purview over a considerable collection of human remains transferred from the Army Medical Museum; he was offered laboratory and office space to begin his work.[9] The collection he was assigned to curate was already impressive—drawers and shelves filled with rare skeletons from around the world—but its physical disarray on the museum's shelves echoed the intellectual uncertainty surrounding the study of human remains. Overall, however, Hrdlička was impressed with the collections at the Smithsonian. But he noted that despite the wealth of American Indian crania, the museum still lacked significant collections of European American or African American remains—a discrepancy he hoped to correct by simply collecting more skeletons. If the most critical question facing American anthropology of the early twentieth century was the unique racial mixture of the Americas, it was perceived as logical to collect all kinds of skeletons and not just those of indigenous peoples.

Hrdlička, who not only frequently possessed strong opinions but maintained a steady desire throughout his life to openly express them, wanted specifically to find skeletons possessing clues about the early arrival of humans in North America. He was stubborn, impatient, and sometimes even aggressive in achieving this goal. Most people found him intimidating. Despite his avowed interest in the question of the prehistoric arrival of humans in the Americas, his absolute top priority for the human remains collections at the Smithsonian when he arrived at the museum continued to be the racial question. In 1903, he noted, "The problems of the utmost interest in physical anthropology in this country concern the course of development in the negro and Indian children."[10] Of equal interest to Hrdlička were the third-generation children of American immigrants, whom he hoped to use for a case study examining how adapting to the environment of the American continent shaped human anatomical development. One year later, in his published guide instructing outside researchers and potential donors how to best collect and preserve human remains, he noted that these kinds of collections were intended to study the "variations in the human body and all its parts, . . . particularly the differences of such variations in the races, tribes, families, and other well-defined groups of humanity."[11] Despite the increasingly vocal calls for the preservation of antiquities of the period, the guide said little regarding the ethics of collecting human bodies. The primary problem for museums was not the ethical acquisition of remains but the questions of what exactly to collect, how to obtain it, and the best storeroom methods for the skeletons that were successfully procured. These were not just simple questions of storage, though best methods for long-term storage of bones were discussed in some depth; the manner in which bones were preserved and organized also represented classificatory frameworks ascribed by scientists. The age of skeletons held some presumed significance, but the *racial origins* of collections was of utmost importance in keeping bone rooms organized.

Museums concerned with physical anthropology and comparative anatomy made the practice of acquiring human remains, especially bones and mummified bodies, a priority in order to establish and build permanent skeletal collections. Measurements of the living were useful, but the bones of the dead were easier to study and restudy as required—locked away within rooms far removed from the public exhibits in most museums. Preserving collections of bones became especially important, as anthropologists and anatomists had idiosyncratic measurement techniques that made use of different tools, resulting in data both unreliable and lacking the uniformity craved by scientists. At times, these differences became heated points of contention among museum scientists. Hrdlička wrote in his introductory guide to the collections,

> The living are examined, measured, photographed, and cast, either in laboratories or in the field; but dead bodies or any parts of them can be studied or prepared for study, demonstration, or exhibition only in laboratories

specially fitted for the purpose. They must be gathered from hospitals, morgues, and dissecting rooms; cleaned, catalogued, and numbered; and then properly stored for preservation, reference, or further investigation. It is plain that such materials can be utilized profitably only in large institutions which can furnish and maintain laboratories, give proper care to the material, and have space for exhibition and storage.[12]

Most skeletal remains added to museum collections in the United States arrived from the field, collected somewhere along a blurred spectrum between early archaeology, looting, and grave robbing. Rather than promoting the collection of recently dissected bodies at the expense of archaeology, Hrdlička hoped to promote a global, systematic collection of human remains representing the diversity of humankind.[13] Bones from dissected human cadavers used at medical schools also ended up in natural history museums, but these bodies and body parts more commonly went to medical museums.

In January 1900, in the midst of the Philippine-American War, the *Chicago Daily Tribune* lauded the government's acquisitioning important skeletons for science, in an article featuring the headline "Uncle Sam's New Islands"; the subheading continued, "Skeletons of Black Dwarfs Brought Back from the Philippines—Almost Missing Link." Although the article expresses a scientific interest in the prehistory and origins of the people in the region, its language is avowedly racial and hierarchical, describing skeletons acquired on behalf of the U.S. government and en route to the Smithsonian as "two skeletons of dwarfs belonging to the curios aboriginal race of that archipelago known as Aetas, or more familiarly as 'Little Niggers.' They are the first of their kind ever fetched to this country, and are considered highly interesting, inasmuch as they represent the lowest existing type of osseous framework."[14]

In late 1905, a young anthropologist named Samuel A. Barrett, traveling along the Putah Creek in Northern California, came across a burial ground. Barrett was a rising scholar who was soon to reach prominence studying the ethnography of California Indians and directing the Milwaukee Public Museum. After examining the site, he wrote, "owing to the more recent relic hunters' visits to this site, there is at present evidently very little to be found in the way of scientific materials." Looters walked away with beads, arrowheads, or whatever other objects—of either real or perceived value—they could get their hands on. Nevertheless, Barrett noted, "Little or no attention has ever been paid by anyone evidently to the taking of skeletons or parts of the skeletons, although there are reported two or three skulls now in the possession of white people of the vicinity." He concluded, "Usually, however, the bones have either been cast aside on the surface or have been thrown back promiscuously into the pits as dug."[15] Barrett, like many other young scholars working to collect all kinds of objects for museums, sent three cases of human bones to the museum at the University of California. His experience, like Densmore's less than a decade later, became a common feature of fieldwork among individuals aware of the existence of the relatively new physical anthropology or anatomical bone collections in museums. Scholars connected to museums were well aware at the turn of the century exactly which museums were actively collecting and competing for the now valuable scientific resources in rare human skeletons. Racialized perceptions of living populations—cast as either primitive or vanishing—shaped the perceived value of associated human remains in terms of both behind-the-scenes research and planned exhibitions. Those who worked in the field, therefore, would collect and submit remains to museums, even if they were only loosely affiliated with the project of building bone collections. Looters made the work problematic, tearing apart graves hurriedly—whisking whatever appeared most valuable to the black market. Skulls or other pieces of skeletons were quite commonly collected as relics, and though they rarely possessed anywhere close to the monetary value in looters' soaring imaginations, they were stolen anyway. Museums have long possessed official policies of avoiding the acquisition of looted material; however, the provenance information—or the documentation of each museum object dating back to its original creator (in this case, the deceased)—was typically sparse in this era. Sometimes, looters were brazen enough to vividly detail the narratives of their participation in episodes of grave robbing, and often museums of both natural history and medicine simply accepted these specimens regardless. Skeletons occasionally took a circuitous and even mysterious route to the bone room.

Hrdlička's views on collection priorities grew from his positions on anthropological methodology. Human remains collections, he maintained, were valued for their ability to provide scientists with a large and relatively stable library for observation. He added, however, that bone rooms should *not* serve as pools for *statistical* analysis, deriding such results as overly complex. Certainly, Hrdlička collected data and measurements from both human remains in museums and the living subjects he used for a series of physical measurements now known as *anthropometry*. His version of the science entailed collecting only the simplest of data, though others invented and lobbied for more complex and detailed calculations to be drawn from living indigenous people who were willing to sit down long enough to have the shape of their heads measured and remeasured—such use of more complex statistical methods was, in Hrdlička's view, pedantic and unnecessary.[16] Others added additional layers of theoretical sophistication onto their ideas of racial classification, but the Smithsonian's most prominent scientist in the field routinely rejected them. Despite criticism of certain racial theories, the Smithsonian was eager to draw fundamental observations from them. Skulls, in particular, Hrdlička argued, "preserve the zoological as well as the racial characteristics of the individual, and also the general form and size of by far the most important human organ, the brain."[17] Skulls, unlike certain other parts of the human skeleton, were also sturdy and withstood many of the destructive challenges of nature that led to the decay or destruction of soft tissue. Even in the final stages of bodily decay, under certain conditions, a well-preserved skull and even tufts of hair can withstand the many destructive natural forces wreaking havoc on human flesh, connective tissue, and muscle. The most important feature of the skull—indeed, the one around which everything in the bone room was organized—was its *racial* origin.

Now settled in at the Smithsonian, Hrdlička proposed small and varied exhibits on physical anthropology. The ideas he offered for how best to go about staging exhibits mirrored his ideas about how to organize the collection behind the scenes. The proposed exhibits were intended to show both normal and abnormal human variation, and the scientist sought to display bones of both white and nonwhite individuals, with a particular focus on crania of American Indians. Native American skeletons by this point constituted the largest portions of the collection. Hrdlička also proposed an exhibit space where scientists working in a laboratory, examining both human remains and ethnological objects, would carry on their research in front of the public. Finally, Hrdlička wrote dense scientific papers for distribution to visitors as they viewed exhibits.[18] Although many of Hrdlička's proposals for displays never came to fruition, his position as the curator of the largest collection of remains granted him influence in shaping many outside exhibitions where skeletons *were* displayed in museum contexts. Furthermore, the involvement of the organization of behind-the-scenes storage and care in discussions of the best methods for display points to underlying theories about the utility of human skeletal remains in understanding race. Despite Hrdlička's failure to immediately see these proposals forward to actual Smithsonian exhibits, the proposals point to the centrality of race in the array of ideas justifying the significance of this collection. Theories about racial classification at the turn of the century, in other words, were mapped onto human remains collections both behind the scenes and in major proposals for exhibition.

The first set of displays proposed by Hrdlička for the Smithsonian compared humans with other mammals. This was followed by a brief display on human evolution that featured "representation in casts or charts of what is known in this respect." Exactly how Hrdlička might have chosen to display his ideas about human evolution remained unclear, however; the limited space he wanted for this section suggests his lack of concern with the subject compared to other themes. Moving quickly from evolution, the displays that followed featured human bone specimens that showed human growth and development from child to adult. This was followed by cases showing then-current ideas on racial difference, illustrated through the display of human brains and skeletons, as well as photographs of living individuals and other specimens. The displays closed with examples of deformed bodies and a series on bodily modifications, such as tattooing or piercing.[19]

Hrdlička's initial plans for displays at the Smithsonian reflected common exhibition practices during the second half of the nineteenth century—displays on racial and biological differences between human populations—while downplaying the use (or potential use) of remains for the study of evolution or prehistory. The proposed displays also featured representative selections of the human remains already in Smithsonian collections—a modest collection of human brains and skeletons intended to represent the races of humankind as understood through racial classification theories of the era. Race took precedence over prehistory, despite Hrdlička's own growing interest in the question of the peopling of the Americas. Simply stated, race remained the dominant paradigm through which museum collections of human remains were understood at the turn of the century. This was true both in the practice of physically organizing and arranging the collection and in the imagination of curators hoping for major new exhibitions. The Smithsonian struggled to find funding for the ambitious exhibitions Hrdlička wanted, but museum leaders continued to encourage him to acquire new collections within the framework of possible future exhibitions. Ironically, despite his enormous influence in organizing, researching, and displaying human remains around the country and internationally, Hrdlička never convinced administrators within his own museum to fulfill his plans for a large-scale display on the science of humankind.

By 1906, the Smithsonian's burgeoning collection of human remains was in disarray. Instead of doing new fieldwork searching out bones, Hrdlička begrudgingly stayed behind at the museum as he started reorganizing the already sizable skeletal collection at the museum. By the end of his project of recataloguing, Hrdlička counted about eight thousand skeletons. The museum also maintained a small but growing collection of human brains preserved in fluid, as well as a vast number of animal specimens collected for comparative purposes.[20] The foundation for systematic expansion had been laid.

SKELETONS AND THE AMERICAN ANTIQUITIES ACT OF 1906

Well before the end of the nineteenth century, serious researchers had come to a collective understanding: significant discoveries of human remains were managed poorly on a national scale. In 1887, Washington Matthews of the U.S. Army took over an expedition for the legendary anthropologist and adventurer Frank Hamilton Cushing after Cushing fell ill. Matthews was horrified to observe the state of important human remains littered throughout the American West. One account notes, "[Matthews] found that no attention had been paid to the collection or preservation of human bones, which were extremely fragile, crumbling to dust upon a touch, and which had been thrown about and trampled under foot by curious visitors, so that but little remained of value from the work which had been previously done."[21] Matthews took for granted the importance of certain remains, knowing even better than most anthropologists of his era the intellectual desires for skeletons—especially those of American Indians—for comparative anatomy or racial science. Since the majority of Anglo-Americans accepted the idea that the American Indian was, in fact, vanishing, it is unsurprising that Matthews was horrified to find naturally decaying skeletons—bones upended and exposed by looters—left to dust outside the protection of museum walls.

Other scientists, too, had grown concerned about looting from historic sites around the country, specifically those in the American West.[22] Rediscovering striking archaeological sites in the American Southwest caused a stir both in the academic community and with writers who wrote for popular audiences.[23] The territory of the United States, many were starting to realize, had a deep and rich history extending centuries before European contact. New discoveries in the American West suddenly seemed important to the growing investigation into ancient history in the Americas.

Early versions of the American Antiquities Act struggled to articulate the place of the national museum, the Smithsonian Institution, as the guardian for antiquities of national significance, in relation to other museums around the country. Earlier drafts of the bill would have had a major impact on state and local museums—whose modest bone rooms were allowed to grow—in some states becoming official repositories for accidental discoveries of historic and prehistoric remains. Beyond the issue of certain museums as official repositories for archaeological remains, the exact mechanism for proper treatment of discoveries was unclear. While Hrdlička envisioned the Smithsonian as becoming a leader in the collection and study of human remains, he never lobbied Congress to designate the museum as an official repository for discoveries of bodies on archaeological sites on a national scale. Although Hrdlička probably never wielded enough influence to successfully recommend otherwise, this indirectly allowed for bone rooms at other natural history and anthropology museums—including those in New York, Chicago, and Philadelphia—to continue to grow as archaeologists and private individuals continued to submit their own discoveries of skeletons from archaeological sites to museum collections.

By the middle of the 1890s, newspaper editorials started to echo professional calls for the preservation of antiquities. An editorial appearing in the *New York Herald* in 1896 argued that "ignorant relic hunters" were clearly to blame for the destruction of antiquities and that only congressional action could save rapidly vanishing sites. The article urgently informed readers, "All these invaluable possessions are fast disappearing, simply for lack of proper legislation to protect them." As proof, a growing tourist market had created a demand for ancient artifacts and works of art, which were easily bought and sold throughout the American West. If the government failed to act, the editorial warned, American heritage, in the form of "our heirlooms from the American aborigines," would be unstudied and forever lost.[24]

In 1904, Franz Boas, then working for the American Museum of Natural History in New York, corresponded with Alice Fletcher, an American ethnologist who became active in the movement to preserve antiquities, writing letters of support even while working in the field. She wrote to Boas, "I understand that the bill lodging the control over all Indian remains in the secretary of the Smithsonian Institution has been introduced. . . . I understand that the bill was dead before it was introduced, but we do not want to take any chances."[25] For Boas, as with others concerned with the impending legislation, "remains" and "antiquities" included everything from massive stone structures to small arrowheads. These terms *also* represented archaeological human remains found on public lands. The mission of collecting human remains both synchronized with and actually advanced the arguments of those who were lobbying for the American Antiquities Act. While Boas was in favor of preserving these discoveries and subjecting them to professional, archaeological scrutiny, he was fervently opposed to the idea that the Smithsonian would be declared the official repository for *all* archaeological skeletal remains discovered on public lands in the United States. Advocates in favor of legislating burial mounds, cliff dwellings, and other discoveries believed to possess historical significance on public lands represented museums in New York and Washington, DC, as well as including anthropologists writing from the field. The chorus was strong enough to push Congress to finally pass the American Antiquities Act of 1906, which ultimately was signed into law by Theodore Roosevelt.

More than a generation before the passing of the American Antiquities Act, looters, amateur archaeologists, and tourists started ravaging sites of ancient human occupation in the American Southwest. The market for antiquities formed a series of steady streams for museums in the United States. Artifacts in private collections were frequently donated to museums (later generations often failed to possess the esoteric drive for collecting mummies or skulls for display above the fireplace). With the rediscovery of Mesa Verde—a vast and complex series of stone structures in Colorado—the Wetherill brothers marked the most prominent of a series of finds in the late nineteenth-century American West.[26] Major discoveries at spectacular places like Mesa Verde, which soon became a national park, became examples that punctuated the countless other smaller incidents of the removal and sale of artifacts—including human remains—occurring at the same time in the United States.[27] Starting around the turn of the century, archaeological sites in the American West became popular tourist destinations, drawing hordes of elites from the East Coast and even a few curious travelers from Europe.[28] Whereas a generation of elites in the United States had visited museums and fairs to view the striking remnants of ancient North American civilizations, the expansion of the railroad allowed a new generation of tourists to see the West firsthand. Many individuals simply could not resist the temptation to take an ancient artifact home with them as a souvenir, including human remains found in American West. Displays of skulls were commonplace in rural homesteads—a symbol of life, death, and the exotic Native American or pioneer history. Collecting and displaying skulls that had been found on farmsteads seemed almost fitting for many would-be collectors. The media and works of popular fiction popularized the notion of the West as a site for long-forgotten civilizations, cowboys, pioneers, and Native Americans.[29]

In 1905, a feature article in the *Los Angeles Herald* proclaimed, "[The Southwest] has made a lasting impression on all students, for it is to them what Egypt and its ruins are to Europe. A land of antiquity, rich with the remains of an almost forgotten past. A land enveloped in a cloak of dust with which kindly nature has hermetically sealed her treasures."[30] This sense of mysticism about the treasures hidden in the American West helped fuel museum administrators' desire to rescue human remains from that "cloak of dust," and professional associations responded by forming committees to push for the passage of an act to prevent further looting from important sites.[31]

The final version of the American Antiquities Act legally protected antiquities found on lands held in public domain and instituted penalties through fines. Subsequently, lands with important archaeological, paleontological, or historical material were eligible to become national monuments, thus providing federal protection against damage. This protection, however, was limited. Archaeological *objects*, rather than human remains, were the priority made explicit in the language of the bill. Several versions of the bill had worked their way through the House and Senate, with some versions even providing for the specific protection of "any aboriginal structure or grave on the public lands of the United States."[32] Although the work of professional organizations and the early draft proposals viewed by Congress pointed to "cemeteries, graves, [and] mounds,"[33] the final version of the bill notably fails to identify graves and cemeteries specifically. Before the twentieth century, the language of preservation often lumped together human remains and archaeological objects, making equivalent, in practice, the preservation of stone tools and naturally mummified remains. Not only was the language left vague, but without a robust service physically protecting the sites—such as the National Park Service does today—the law initially provided only an easily penetrated shield around historically significant sites.

Early in the development of the American Antiquities Act, Congress turned to experts. According to the official Smithsonian report, William Henry Holmes, the chief of the Bureau of American Ethnology "was called upon to assist in formulating the uniform rules and regulations required by the Departments of the Interior, Agriculture, and War in carrying out the provisions of the law for the preservation of antiquities, to pass upon various applications for permits to explore among the antiquities of the public domain, and to furnish data needful in the selection of archaeological sites to be set aside as national monuments."[34] Holmes, in turn, supported the efforts of a politically shrewd young archaeologist from New Mexico named Edgar Lee Hewett.[35] Ideas about collecting and research relevant to the Smithsonian's anthropological collections therefore flowed directly from the museum to Capitol Hill, informing the final specifications of the bill.

In 1904, Hewett, a scholar skilled at working with government officials, bureaucrats, and scientists alike, launched a review of the American Indian antiquities of the American West.[36] Hewett's lobbying included letter writing and the publication of pamphlets that were circulated to concerned anthropologists and archaeologists (including those working in the field, like Alice Fletcher), as well as to politicians in Washington.[37] His work advised Congress of the various problems related to preservation in the region and helped shape the final language of the bill.[38] Hewett worked to navigate tensions between the Office of the Interior and the Smithsonian Institution, as the two agencies maintained contesting visions over the nature of the bill.[39] Congressional reports on the proposed bill added, "It provides that any person who shall appropriate, excavate, injure, or destroy any historic or prehistoric ruin or monument, or any object of antiquity, situated on lands owned or controlled by the Government of the United States without having . . . permission . . . shall, upon conviction be fined in a sum of not more than $500 or be imprisoned for a period of not more than ninety days."[40]

The language of the bill points to the preservation of "antiquities," but it was leveraged in practice by Theodore Roosevelt to protect sites of both historic and *environmental* significance. Between 1906 and 1908, historic sites including Montezuma Castle in Arizona, Chaco Canyon, and the Gila cliff dwellings in New Mexico were approved for protection under the act. Over the same span of time, Devil's Tower in Wyoming, the Muir Woods in California, and the Grand Canyon in Arizona—all sites known for remarkable environmental significance—became national monuments under the same provisions.[41] When the discovery of ancient or mysterious bodies resulted in their high-profile removal to museums for further study and display, the story had the unintended consequence of lifting environmental preservation efforts in the American West. The organic political connection between federal governance of ancient graves and environmental preservation efforts went almost completely unnoticed as these events took place.

After the passage of the law in 1906, Congress appropriated $3,000 for two years for the "excavation, repair, and preservation" of the Casa Grande Ruin in Arizona.[42] Several other ancient monuments in the Southwest soon followed. The allotment for actual protection of historical sites and monuments was grossly inadequate, but it was a start; Congress gradually appropriated more funds to protect and preserve public lands of environmental and historical significance. In subsequent decades, the National Park Service built its own collection of materials—including human remains—discovered on federal lands. Museums, and archaeologists working on their behalf, now needed to apply for permits to collect archaeological material—including human remains—from federally owned sites. While the laws worked to protect sites from looting, they generally did not prohibit scientists and explorers who wished to deposit their discoveries at museums of natural history or anthropology.

Within a decade of the passage of the act, the U.S. Railroad Administration, the National Parks Service, and the Denver and Rio Grande Railroad began to jointly publish maps and pamphlets promoting tourism to the recently preserved archaeological sites of the American Southwest. A promotional pamphlet published by the National Park Service sometime after 1916 features an introductory quote from the secretary of the interior, Franklin Knight Lane. Lane assured potential visitors that "Uncle Sam asks you to be his guest" and that the parks were "the playgrounds of the people." Following this was a more complete description of the sites around Mesa Verde National Park, including a series of photographs and maps of the park intended to orient visitors geographically. Pictures featured in the promotional pamphlet were taken by George L. Beam, a noted southwestern photographer. The last photograph in the collection featured a human skull and a series of long bones surrounded by a group of twelve impeccably preserved ancient clay jars. The caption notes the rarity of the jars but makes no mention of the human skeletons, despite their unmistakable prominence in the promotional photography.[43]

Although the American Antiquities Act was vague with regard to its legal guidance for the treatment of human remains found on archaeological sites, it did represent a step in the direction of the professionalization of archaeology in the United States. Just as Hrdlička was organizing his existing collections of human remains, legislators were establishing rules for antiquities as they were to be collected in the field. The American Antiquities Act not only had direct and obvious consequences for archaeology in North America; it also enacted a series of far less obvious consequences for general environmental and historical preservation in the American West. The popular presentation of rare and ancient skeletal remains—by scientists, government agencies, and the media—played a significant role in shaping ideas about attempts to collect bodies for the study of race and human history in this era.

WORLD WAR I AND "WAR OPPORTUNITY"

Warfare created numerous problems for those who wished to collect and study human remains. Wartime, however, also created new opportunities to collect skeletons. Throughout the nineteenth century, scholars exploited war to collect human remains across North America. Samuel George Morton's collection of skulls included crania collected in the wake the Battle of Lake Okeechobee (1837) of the Seminole War in Florida, as well as the Mexican-American War (1846–1848).[44] Later, through the Civil War and the ensuing Indian Wars, medical doctors leveraged conflicts to build collections at the Army Medical Museum. When the First World War began, shipments of remains from Europe slowed and nearly halted. Scholars who had previously been allowed access to original materials in European museums were now forced to stay at home.

Those who were involved in the planning of the anthropological exhibitions at the 1915 San Diego world's fair anticipated potential problems brought on by wartime in the planning stages for the fair. Aleš Hrdlička wrote as much to the archaeologist Edgar

Hewett: "The terrible conditions in Europe I am very much afraid will interfere with us, though to what extent I am not able to say."[45] By August 1914, Hrdlička noted in letters related to the planning of the fair, "The European war, I am now very apprehensive, will strike us heavily."[46] Indeed, the once-steady flow of remains arriving at museums in the United States from abroad temporarily slowed during the war.

Despite these problems, the First World War created numerous, and largely unexpected, opportunities to collect remains. The Smithsonian, in particular, started to receive an increased number of offers for *sale* of specimens from Europe. Hrdlička, in fact, referred to the offer of eight gorilla skulls at the prewar price of a single skull as "another rare War opportunity."[47] Throughout the course of the conflict, Smithsonian staff members were asked to contribute to the war effort in a number of ways. In particular, curatorial staff provided the government and military with information perceived to be important to fighting overseas. The physical anthropology division, in particular, "furnished a large amount of information on racial questions" to the National Research Council and the Army and Navy Intelligence Bureaus.[48]

The massive effort undertaken by large armies to dig trenches during the European war also uncovered artifacts and examples of archaeological remains. Hrdlička wrote to his supervisor and friend, Holmes, to explain the need to contact the army to provide detailed instructions, a protocol, for soldiers who accidentally uncovered ancient skeletons:

> France was a home of Early man throughout a large part of the period of his evolution. In many parts of France archaeological and skeletal remains of ancient men have been discovered, and many doubtless lie yet in the soil. It will not be long before our Army will be making many trenches and dugouts in France, and it is more than

probable that during this work more or less ancient human remains will be repeatedly discovered. Such discoveries have already been made in the trenches by the French themselves, by the English and also by the Germans. As the scientific value of the objects recovered may be very great, it seems indicated that proper steps be taken for their preservation."[49]

Hrdlička suggested that army officers be made aware of the fact that their activities may uncover skeletons of prehistoric significance. He suggested that discoveries might go to the French government or to the Smithsonian. As if returning to the letter later in order to underscore his point, he scratched a note in pencil next to the typewritten letter: "The main thing is to save them for science."[50]

The war presented an opportunity for the acquisition of new human remains and new animal skeletons for comparative purposes, but it also presented an even larger opportunity for new anthropometric measurements. Hrdlička frequently called on the army to measure the bodies of incoming recruits for science, recording simple information about the body and heritage to complement studies of stable collections of human remains. Hrdlička eventually trained army officers to conduct exact measurements of incoming recruits and soldiers who were sent to hospitals.[51]

Despite political upheaval and travel restrictions, museums in the United States, unlike their European counterparts, largely benefited from the outbreak of World War I. Museums in the United States, again quite unlike their European counterparts, did not experience the full effects of the economic and material devastation and chaos of war. Despite a lack of funds, American scientists were generally afforded quiet and peaceful time to study collections. Still, thoughts of young men digging trenches and accidentally discovering significant skeletal remains kept many scientists awake at night during the war. Meanwhile, young men worried more about surviving war than collecting ancient skeletons, mostly ignoring requests to be mindful of old bones.

THE MATTER OF ISHI'S BRAIN

Just before U.S. entry into the war, anthropologists, historians, and the public became enthralled by a story in California. In August 1911, a lone Yahi Indian man came out of hiding near the town of Oroville. The man was emaciated and visibly confused. Not knowing what to do with the man (a "wild" Indian had not been encountered for years), the townspeople contacted the local sheriff, who brought him to the nearby jail. Eventually, the anthropologist Alfred Kroeber of the University of California, Berkeley, was contacted to identify the mysterious individual. Kroeber had been a student of Franz Boas at Columbia before settling in California. Both Kroeber and his former mentor trained influential anthropologists over the course of the ensuing decades, shaping the field through scholarly publications and widely used textbooks.[52] Anthropologists training under Kroeber were given a broad anthropological education in the traditional four-fields scheme—linguistics, ethnography, archaeology, and physical anthropology—yet the majority of his students engaged in research related to the ethnography of California Indians. Despite his typically professional and emotionally removed nature, students and faculty in California found Kroeber a capable intellectual leader, many even mimicking his trademark beard, which came to a clean point at the chin. Kroeber was generally a calm and thoughtful intellectual who remained engaged in the necessary campus affairs, though he found himself occasionally irritated by the legendary Berkeley bureaucracy. Interested in the claim of a discovery of a "wild man" in California, Kroeber traveled to Oroville and attempted to speak with the man using pieces of other California Indian languages. Eventually, the man was determined to be Yahi, and he was called

Ishi, which means simply "man" in the Yana language. Kroeber had him brought to the University of California Museum of Anthropology in San Francisco, where Kroeber cared for and studied Ishi as he continued to mentor, teach, and curate in the same building.[53] Scholars and, especially, members of the media—considered what the so-called discovery of the "last wild Indian" might mean for scientific studies of both race and primitive man.

Most anthropologists and historians of the American West today are familiar with the story of Ishi—America's last "wild" Indian—and attempts to interpret the meaning of his life have been both numerous and diverse in their conclusions. The story of Ishi's physical remains was intimately linked to the framework of collecting human remains to study race. Ishi's "discovery" and subsequent death took place when scholars were actively trying to record every scrap of information about indigenous races, attempting to preserve words, traditions, and—when the time came—cadavers.

Born around 1860, Ishi lived with a small band of his tribe that managed to escape the attention of surrounding whites who were migrating by the hundreds of thousands to California. Over time, Ishi's band dwindled until he was the final remaining member. He continued to conceal himself until 1911. Almost immediately after Ishi was "discovered" in Northern California, those who were interested in the history and culture of the American Indian began studying him, trying to salvage every scrap of information about his culture through interviews, questions, and slow dialogue. Residing at the University of California Museum of Anthropology, Ishi appeared to experience an outwardly content life. Ishi even became friends with a medical doctor at a nearby medical college, and the two frequently practiced archery in the lawn between the museum and the hospital. His celebrity and affable personality became fodder for tabloid media, and he became well known around San Francisco. Although he periodically toured the medical school's facilities and was fascinated by surgery, he possessed a strong culturally based revulsion to the bodies of the dead. In photographs, he appears either strong and silent, depicted in a long tradition of the conceptualization of the "noble Indian," or smiling happily with his new life and friends.

In 1916, while Kroeber was traveling abroad, Ishi died of tuberculosis. Staff at the University of California wrote the anthropologist, notifying him of Ishi's imminent death. Kroeber responded by demanding that no autopsy be conducted. Because of Ishi's complex attitudes toward the bodies of the dead, Kroeber instructed his staff, "We propose to stand by our friends. If there is any talk of the interests of the science, then say for me that science can go to hell."[54] The letter, however, arrived after an autopsy had already been conducted. When Kroeber arrived back in Berkeley, he returned to the news that while Ishi's body had been cremated in accordance with his wishes, the brain had been preserved. Though the University of California Museum of Anthropology possessed a collection of skeletal remains and mummies, it did not maintain a brain collection. Kroeber eventually decided to contact officials at the Smithsonian.

Scholars dispute Kroeber's rationale for sending the brain to the Smithsonian, rather than reburying it. The anthropologist Nancy Scheper-Hughes speculates, "Kroeber's behaviour was an act of disordered mourning. Grief can be expressed in a myriad of inchoate and displayed ways ranging from denial and avoidance, as in the Yahi taboo on speaking the names of the dead to the insistence that the death and loss experience is a minor one." She further notes that Kroeber avoided discussing Ishi following his death and that the subject of Ishi caused Kroeber a great deal of psychological pain.[55]

While Kroeber may have wished to distance himself from the bodily remains of his friend for emotional reasons, the rationale for donating the brain to the Smithsonian was also clearly academic. When the Smithsonian learned of Kroeber's interest in submitting the brain of the well-known California Indian to their growing brain collection, officials were enthusiastic. What particularly excited them was the prospect of acquiring the brain of an individual about whom so much was already known through efforts to salvage his culture, through ethnography, before his ultimate demise. Ideas linking the development of the brain and cultural habits were still prevalent, and Hrdlička noted with keen interest that Ishi's case was special, as he had been studied during his life by a number of highly regarded anthropologists.

Hrdlička had studied the anatomy of the brains of indigenous peoples before, publishing one account of an autopsy of a brain of an adult male Eskimo named Kishu in the journal *American Anthropologist*. Kishu, like Ishi, had succumbed to tuberculosis while in the care of anthropologists and medical professionals.[56] Kishu perished in New York City at Bellevue Hospital, after being measured, photographed, and displayed during his own life at the American Museum of Natural History.[57] By the time of Hrdlička's writing in 1901, Kishu's brain had already been added to the anatomy collections at Columbia University.[58] Hrdlička happened to be in New York City at the time, and he was given the opportunity to examine the brain alongside other medical professionals in the area. Hrdlička concluded that Kishu's measurements did not make him "racially exceptional." In other words, he fit within a perceived racial typology when his body was compared with other individuals from the same region. Hrdlička compared Kishu's brain to whites', arguing that certain parts of the brain were more or less developed than that of the average white male.[59] On the whole, however, Hrdlička noted, "this Eskimo brain is heavier and larger than the average brain of white men of similar stature."[60] Whereas earlier studies of human remains had focused their attention on the capacity and shape of the skull, Hrdlička was starting to put forward the argument that one might compare the differences of "the brain in different individuals and especially in different races."[61] If

Kroeber wanted to divest himself of Ishi's brain for a combination of personal and intellectual reasons, submitting the brain to the Smithsonian must have seemed reasonable. Hrdlička moved to the Smithsonian in 1903, and his reputation for studying and collecting brains, in addition to a much larger collection of skeletal materials, no doubt followed him to Washington, DC.[62]

A series of letters exchanged between December 1916 and January 1917 among Aleš Hrdlička, the curator of physical anthropology; William Henry Holmes, the head of the Smithsonian's anthropology department; and Alfred Kroeber point to other nuances. In writing to Hrdlička, Kroeber stated plainly, "I find that at Ishi's death last spring his brain was removed and preserved. There is no one here who can put it to scientific use. If you wish it, I shall be glad to deposit it in the National Museum Collection."[63] Although Kroeber's museum was embarking on the construction of the largest human skeletal collection in North America west of the Mississippi, the strength of the materials was to be skeletons from California and the Great Basin, along with a smattering of mummified remains from the American Southwest. Hrdlička quickly responded to Kroeber, expressing his enthusiasm that the brain might come to the Smithsonian. Ishi's brain added to the museum's growing collection of human brains, which by this time already stood at over two hundred.[64] On December 20, 1916, Hrdlička first wrote to his supervisor, Holmes, on the subject of Ishi. The two were close colleagues and frequently corresponded about official matters. Hrdlička wrote simply, "Prof. A. L. Kroeber of the University of California has kindly offered us, as you will see from the accompanying correspondence, the brain of Ishi, the last survivor of a trip [*sic*] of California Indians. I beg to recommend that direction be sent to him for shipping this specimen by express at our expense. I have already given him instructions as to packing."[65] Several weeks later, Kroeber responded to Hrdlička's official letter of acceptance, asking for instruction as to how, exactly, to ship Ishi's brain. Skeletal remains were one thing, but a soggy ball of flesh and blood was quite another. Kroeber, having been trained as an anthropologist in this era, was certainly familiar with skeletal material, but soft-tissue remains were generally unfamiliar to scholars not trained in medicine. Hrdlička, trained as a medical doctor and having already built a collection of over two hundred brains, was intimately familiar with methods for packing and preserving soft tissue. He responded to Kroeber in some detail, referring to the additional instructions he hoped Holmes would provide: "As to the shipping of Ishi's brain, you will receive in a day or two a communication which will give the exact directions. The brain should be packed in plenty of absorbent cotton saturated with the liquid in which it is preserved, and the whole should be enclosed in a piece of oiled cloth or oiled paper. The package should then be laid in a moderate sized box with a good layer of soft excelsior all around it. In that way it will doubtless reach us in good condition."[66] In a final step intended to transfer ownership of Ishi's brain from Kroeber and the University of California Museum of Anthropology to the Smithsonian, Kroeber wrote to the assistant secretary of the national museum, "Replying to your favor of December 30 I would state that we are shipping you the brain of Ishi per your directions, express, consigned to the United States National Museum. The packing follows the directions of Dr. Hrdlička."[67]

Though those few lines alone would have been enough to transfer the title of Ishi's remains from the University of California to the Smithsonian, Kroeber was a museum professional and an avid collector of anthropological material.[68] As such, he was intimately familiar with the problem of provenance. Many of the objects in his own museum, including a number of the skeletal remains in the University of California's growing collection, lacked associated provenance information. This lack of provenance information was a problem shared by other museums around the country, nearly all of which possessed objects collected before the supposed "professionalization" of the discipline, which encouraged anthropologists to carefully record the origin of the objects they collected for museums. Despite the fact that it was taken for granted that natural history museums collected both bones and baskets, most donors frequently turned over specimens without much additional information. The fact that many objects were catalogued with little associated data frequently led to problems in attempting to conduct modern science on collections. Kroeber understandably, then, added the following information to his letter:

> I add the following for your records: Ishi belonged to the southern most of four divisions of the Yana stock in north-central California. In lieu of a proper tribal name we are adopting the designation Yahi for the group in publications concerning them; this being Ishi's word for people. The habitat of the group was on Mill and Deer Creeks, Tehama County. The tribe was virtually exterminated about 1865 by the settlers. Four or five survivors maintained a precarious existence in the hills from that time until 1906 when they were rediscovered by accident. During the following three years they all perished except Ishi, who toward the end of August, 1911 was found near Oroville in Butte County where he had wandered from his native territory. From that time until his death, March 23, 1916, he lived at the University of California Anthropological Museum. The cause of his death was tuberculosis. I estimated his age at the time of death at 55 years. Most medical men who have examined him are inclined to put the figure somewhat lower. I have a few bodily measurements of Ishi, which I shall forward to Dr. Hrdlička.[69]

Kroeber was not entirely silent on the subject of Ishi following his friend's death in 1916. In his professional writings, Kroeber was rarely given to emotion, so his use of purely "scientific" language is hardly surprising. Although the drive behind his submission of Ishi's brain to the Smithsonian may have been due in part to an emotional response to his friend's death, it is clear that Kroeber came to the realization that if Ishi's brain had already been removed from the rest of his earthly remains, it should be deposited into a collection where it was likely to be most utilized. The University of California was not as renowned for its collection of human brains as was the Smithsonian, and the logical thing to do, therefore, was to offer the brain to Hrdlička. Science, as it turned out, was not cast off to hell, as Kroeber initially wished it to be. The vision that the Smithsonian maintained for brain collections, detailing scientific information about race through close observation and study of particular body parts—including the brain—speaks to the power of the idea that certain bodies belonged in museums in order to more fully understand the races of humankind. Despite the pervasive nature of this idea, the moment of rapidly collecting, observing, and displaying bodies around the central tenet of racial classification was reaching its crest. Ishi not only represented the last of his tribe; he also represents the near conclusion of a moment of creating physical anthropology collections in natural history museums primarily for the goal of improving scientific understandings of racial classifications.

SKELETONS AND THE EARLY STUDY OF PREHISTORIC MAN IN THE AMERICAS

As museums continued to acquire remains intended for the comparative study of race, scholars began to turn at least some of their attention to more historical questions; an interest in studying the bones of the ancient dead became increasingly central in physical anthropology in both the United States and abroad. Actual discoveries augmented an existing desire to find a "missing link" between modern humans and their forbears. Hrdlička, in explaining his desire to exchange remains with a museum in Siberia, noted that he was, "at present, occupied with investigations of the physical variations which exists between the Siberian and the American natives, and [he] was surprised by the numerous resemblances between the two peoples."[70]

The question of how people arrived in the Americas increased in importance for scholars between 1910 and 1920.[71] Although directly related to the attempts to preserve and compare races—as witnessed in the ideas that surrounded the preservation of Ishi's brain—scholars were increasingly adopting the lens of *prehistory* in order to interpret human remains. By 1914, Hrdlička considered discovering the origins of American Indians "the most important task in American Anthropology."[72] Hrdlička had, in fact, been brought to the Smithsonian largely to address the question of how long humans had occupied North America. Holmes worked to persuade him that humans did not have a long history on the continent and that searching for fossilized skeletons of a much deeper origin would therefore be fruitless.[73] Hrdlička based his assumptions about the relatively recent arrival of humans in North America on his firsthand comparisons with prehistoric remains in Europe. T. Dale Stewart later recalled of Hrdlička in his oral history, "He'd seen all the remains of ancient men in the Old World and was well aware that the old ones over there were very different, very primitive looking, and he had seen nothing in his experience here to . . . lead him to believe that there was anything but the Indian type in America."[74]

Hrdlička did not fully separate the task of understanding how humans arrived in North America from the existing project of developing schemes for racial classification based on measuring more recent skeletons. American Indians were a distinct racial type to be understood historically—an idea that some scholars termed *racial history*, phrasing that was about to come into even greater vogue over ensuing decades. Hrdlička explained in a letter that research of the most pressing importance was to come from Asia, but "research of more local importance which, however, as our classification of the Indian types and subtypes, becomes more and more necessary, is the physical examination of the rapidly disappearing full-blood remains of a number of U.S. tribes."[75] Scholars therefore should look to Siberia, Alaska, and the American West for answers, and priority would be given to attempts to collect skeletons from these regions.

Hrdlička viewed prehistory and human evolution as global questions (with certain regions expected to reveal the most exciting and unique results). Collecting and researching specimens from Europe and Asia as well as those from the Americas was thought to be most crucial in attempts to address pressing questions in racial history.[76] Collections brought together from North America alone would simply not suffice in understanding the peopling of the Americas. In 1913, for instance, the Smithsonian announced new skeletal collections from Mongolia as especially significant—implying that these remains could potentially be useful for studying this major question in future research. Perhaps only coincidence but more likely a signal of how the museum viewed the newly acquired set of remains, the Smithsonian mentioned the acquisition of the Mongolian skeletons in its annual report immediately following the announcement of the most important acquisition of the year, "The Star-Spangled Banner."[77] As these connected themes—the search for racial history and the construction of racial classification theories—grew into prominence simultaneously in the early decades of the twentieth century, old tensions over how best to organize and maintain physical anthropology collections were being addressed in new ways.

"ENTIRELY IN STORAGE"

Museum curators, collectors, and colonial explorers who shipped remains back to museums in the early decades of the twentieth century often maintained grandiose ideas for future exhibitions as they steadily built collections.[78] Fantasies about extensive museum displays centering on racial history struggled to become reality, and the focus of ongoing debates surrounding the physical practice of curating such remains continued to center on best practices for museum organization more generally. Remains that were not currently being researched or exhibited had to be stored, preserved, and organized according to the dominant anthropological theories of the day. Grouping and organizing a wildly diverse collection of bodies, bones, and body parts became an enormous challenge—and one echoing debates about how to understand the remains in the first place.

Research was built into exhibitions in other, less obvious ways as well. In 1911, when Hrdlička was noting that his next project would focus on living and ancient Pueblo Indians, he explained that he would need comparative measurements including of "white native Americans, preferably those of the third generation, between 28 and 50 years of age." Although he was surrounded on a daily basis by the living examples of European Americans, Hrdlička seems to have, ironically, experienced unexpected difficulty in finding a satisfactory number of whites willing to sit in order to have their heads measured. If the process of measuring skulls were to be a part of the Smithsonian's exhibitions, Hrdlička postulated, then perhaps people would be more willing to be measured. While he may not have known it at the time, removing himself from the process of actually measuring museum visitors was likely a good idea, if the desired result was a greater number of volunteers—Hrdlička's gruff approach to scientists, members of the public, and life in general may have contributed to a general reticence to sitting anywhere near the curmudgeon while he was wielding metal calipers. Hrdlička stated, "There are many visitors every day to the National Museum. If a proper notice were placed in a prominent place in the two buildings, calling the attention of the visitors to our need in the above respect, I believe that a number would be induced to permit me to take the measurements."[79] If only visitors knew that they were contributing to *science*, it was argued, they would be happy to contribute to ongoing racial research.

While the Smithsonian was working toward displaying human remains under the auspices of physical anthropology, not all museums of the era had the space, resources, or desire to display their existing collections of remains. In 1918, Kroeber wrote to Hrdlička from California, explaining simply, "We have no special exhibit relating to Physical Anthropology." While the remains he curated might go on display when relevant to other displays on ethnology or archaeology, Kroeber noted that remains were not *central* to any displays at his museum, which was then based in San Francisco, across the bay from the Berkeley campus. Instead, he wrote, "Our collection of Physical Anthropology is entirely in storage." This was not to say that remains were somehow off-limits. As the curator of a research museum, he noted, "With the exception of the Egyptian material, it is all accessible to students."[80] Regular museum exhibition during the era, therefore, played a limited, if critical, role in the development of physical anthropology. Although scholars like Kroeber had little success in creating permanent displays on lessons of race or human history drawn from skeletal collections, the collections continued to grow, and it was assumed that productive research, and eventual periodic display, would continue to result from the existence of the collections.

ACQUIRING REMAINS ILLICITLY

As the study of remains became more organized, museums continued to acquire remains haphazardly and opportunistically. Although many remains were acquired through archaeological fieldwork, much of which conformed to the new American Antiquities Act, numerous remains continued to be quietly acquired in defiance of national and international laws or cultural values. Understanding the complex origins of these collections is critical to fully comprehend their use for science and anthropology. Although the majority of remains collected by Hrdlička were acquired from archaeological excavations, he did acquire some specimens through purchase. The acquisition of remains from overseas frequently continued to echo the nefarious collecting practices of the Army Medical Museum decades before. Occasionally, remains might be sent to the Smithsonian from a vague source in a foreign country with only brief or incomplete descriptions of the origin of the skeleton. Although not considered ideal, these skeletons were regularly accepted by museums with physical anthropology collections.

During Hrdlička's travels collecting remains for the Panama-California Exposition, local representatives in Peru helped him to acquire over sixty skulls. Hrdlička gladly greased the palms of local representatives for their trouble. Several months later, a Peruvian official who personally assisted Hrdlička reported several drawers of museum specimens as missing. Hrdlička promised to return any skeletal materials he acquired that bore any markings from other museums (predictably, there were none). Hrdlička was suspicious that the Peruvians were sabotaging him, preparing to accuse him of stealing. He described one minister as "jealous of [his] success." Hrdlička added that that he felt "glad to be rid of [him]" after the first portion of his stay in Peru.[81] His conclusion was that the local minister was both corrupt and jealous. It was not surprising that Hrdlička encountered this sort of problem, having explained to his supervisor at the fair just days following the foregoing incident, "There are in Peru two or three first-class archaeological collections, which could be bought and I think, on the quiet, exported."[82]

While Hrdlička's strong personality frequently rubbed others the wrong way, he was remarkably successful at acquiring skeletons from all kinds of sources around the globe. The Smithsonian's physical anthropology collections continued to expand—even if scholars were now forced to confront the dual studies of race and history drawn from the remains. Curators willingly bent their own ethical standards for collections when navigating the treacherous bureaucracies that prevented them from removing remains for science. Taken from a different perspective, however, while local government officials occasionally displayed instances of systematic corruption, local indigenous groups were often powerless to stop the ransacking of graves. The archaeologists and physical anthropologists may have followed the looters, but for some indigenous people, the result was the same. Burials and grave goods were critical materials for archaeological studies, yet at the same time, they were clearly sacred to most ancestral groups.

Despite increased legislation, the market for the trade in human remains did not entirely cease over the ensuing decades. When the anthropologist T. Dale Stewart traveled to Peru, returning on behalf of the Smithsonian years later in 1941, he wrote to Hrdlička about many of the same corrupt actors involved with the trafficking of burial goods encountered by Hrdlička decades earlier. Recognizing the name of the inspector right away, Hrdlička responded, "He doubtless has hidden many a good thing for eventual disposal to a good bidder." The Smithsonian, however, was not necessarily always outbid for collections of human remains in other countries. Despite Hrdlička's mistrust of Peruvian officials, he asked Stewart, "I wonder if you couldn't get him to give a price of the skull with a gold plate—try confidentially anyway and let me know."[83] Museums undoubtedly far preferred to acquire skeletons or mummies from a known professional—preferably those with academic degrees working on behalf of recognizable institutions such as major national museums. Nevertheless, when skeletons became available through opportunities in war, corruption, and global politics, museums of the early twentieth century were often fairly quick to react to the circumstances at hand, sometimes acquiring thousands of remains for study.

When remains were collected through legal means, a letter of introduction was presented to local government officials from the Smithsonian. As the U.S. national museum was located in the nation's capital, embassies for nearly every country were available for consult regarding additional letters of introduction or permits. These letters intended to make collecting skeletons both legal and unfettered by locals.[84] It is unclear, however, how often collectors working on behalf of museums might have shown or translated the letters to local indigenous people on the ground before collecting human remains. Over the ensuing decades, collectors were forced to confront changing ethical and legal guidelines for acquiring the remains they desired for their collections. At the time, however, such acquisitions were viewed as a benefit to science, serving to diversify the scope of bone collections at major museums across the country.

PLAYING ARCHAEOLOGIST

Bone collectors noticeably influenced other collectors working in North America who normally concerned themselves with gathering language data, recording songs, or collecting other artifacts. Several scholars typically remembered as ethnographers or linguists also engaged in bone collecting. Frances Densmore, broadly considered the mother of ethnomusicology, was an avid collector and recorder of American Indian songs and poetry.[85] While historians have long recognized Densmore's contributions to the study of language and music, archival evidence shows that she also contributed a collection of pottery and, more notably, human remains to the Smithsonian. Recent narratives of the history of anthropology have portrayed a small handful of individuals, including Franz Boas, Aleš Hrdlička, George Dorsey, and Alfred Kroeber, as the *primary* collectors of human remains for museums in the United States. Indeed, these individuals, working with archaeologists and anthropologists from around the globe, contributed to the growth of museum collections of physical anthropology in numerous and striking ways.

Moving away from collections reflecting ideas in physical anthropology, other historical figures in the anthropological community, including Frances Densmore, Alice Fletcher, and John P. Harrington, have often been portrayed as outside the colonial appropriation of human remains, instead being seen as trending toward more sympathetic, if at times less "scientific," portrayals of native cultures.[86] While this was certainly the case in their published ethnographic writings, all three of these figures who are typically associated with ethnography contributed small collections of human remains to museums in the United States. Amateur collectors and donors, too, contributed collections of human remains to museums on an opportunistic basis. The idea that skeletons could contribute to science and anthropology was a deep and powerful idea that was recognized by practitioners—professional and otherwise—of all types of anthropology and archaeology. Frequently, ethnographers collecting human remains understood their assistance in collecting skeletons as contributing to general scientific understandings of racial history, even if they only held a limited understanding of the implications of the theories arising from physical anthropology of the era. Bones and bodies, simply stated, were productively placed in natural history museums and held in the same collections, albeit somewhat tenuously, as arrowheads, canoes, and baskets. Although many members of the public might not have been familiar with the activities taking place within bone rooms, scholars who worked with museums, even in a very limited way, were, in fact, aware that skeletons of indigenous peoples could be productively placed in the museum. The idea of the bone room, in other words, had expanded well beyond the scientists who spent their careers working in them.

Had Densmore simply collected available remains and submitted them to the Smithsonian, historians might be able to read the event as part of her larger project to collect and preserve everything related to American Indian culture. Her eager follow-up letters to Hrdlička, however, indicate that her interest in the bones had not dissipated. In her letters, Densmore's tone reflects a genuine personal interest in understanding the significance of the remains. Certainly, all kinds of collectors working on behalf of museums opportunistically gathered materials believed to be valuable; however, bone collecting by individuals who were typically concerned with material culture or language indicates the broad reach of the practice within the anthropological community throughout the nineteenth and twentieth centuries. While these scholars were primarily concerned with the supposedly rapid disappearance of material culture and language, many were open to collecting—and submitting to museums—human remains of the same indigenous populations they studied. The idea that these were vanishing *races* of humankind was powerful, if ill defined. This pervasive idea resulted in the tangible act of gathering and placing bones or mummified bodies into shipping containers, labeled with the addresses of major museums of natural history, anthropology, and to a lesser extent medicine—a patchwork process of building bone collections through opportunistic acquisition.

Alice Fletcher, a noted ethnologist, was one of the most important early women in American anthropology. Like Densmore, her notoriety stemmed from her ongoing (even physically demanding) fieldwork with indigenous peoples across the United States. While she spent some of her early career studying archaeology under the tutelage of Frederick Ward Putnam at Harvard, she is best remembered for her ethnographic work with the Omaha that spanned a forty-year period. In 1884, however, when she was asked by Putnam to collect measurements of living individuals as well as of skeletal material from the field, a practice more commonly associated with physical anthropology than ethnography, she readily obliged. In a letter written later that spring, she wrote, "A little north of here is the old burial site of the Omaha people called the hill of graves. This is now owned by white people. I know personally some of the owners & will be allowed to put a laborer to digging in these graves for the skulls and skeletons & any other articles found with the body. I cannot tell how many I could secure & forward to you but I think quite a number of skulls & etc. of . . . Omahas."[87] If the letters and her detailed description of burial mounds were any indication, Fletcher showed an aptitude for collecting bodies even if doing so was not in her regular purview as an anthropologist. Fletcher added that in connection with the collection of skeletal material, she might be able to acquire a few photographs of nearby American Indians, together with information about their exact linage. She added, however, "It is not easy to get at the people for such things but owing to my peculiar work now in progress I have a hold I can never get again most likely."[88] Collecting human remains and measurements of the living in order to assist other scientists in projects to understand race and human history, she believed, came secondary to her work on the cultural habits of the Omaha. Although this incident certainly does not represent a long-standing, systematic process of collecting human remains for museums, it demonstrates that Fletcher did collect human remains for other anthropologists interested in observing and arranging them in cluttered bone rooms. Furthermore, the request that she do so indicates the general acceptance of such a practice.

Despite the overarching influence of the notion that certain human skeletons belonged in the museum, many museum leaders concerned with anthropology and culture maintained little desire to compete with already formidable collections in New York, Chicago, Philadelphia, and Washington, DC. Certainly, when Ishi's brain was collected by the University of California, it was determined that the unique brain specimen was better suited for the Smithsonian. On the other hand, administrators at museums like that at the University of California were eager to acquire most kinds of skeletal collections, thinking they could contribute to training and research in physical anthropology through bones rather than brains. Some administrators simply chose not to develop collections for physical anthropology and therefore turned over discoveries of indigenous human remains to the proprietors of bone rooms at other institutions. In 1914, after discovering sixty-eight skeletons in the Delaware River Valley, George Gustov Heye, founder of the Heye Museum in New York, wrote to Aleš Hrdlička requesting that he write a brief article about the skeletons. Heye noted, "I feel the only place to deposit specimens of this kind is with you and if the National Museum have [sic] not already too many skeletons from that region, I would be most glad to donate the entire lot to your Institution."[89] The Heye Museum, which much later became the core collection of the Smithsonian's National Museum of the American Indian, chose not to accession human remains collections for its own museum. Simply stated, smaller museums such as the Heye were simply too busy building their own empires to worry about bones, and instead they occasionally submitted discoveries of skeletons to the museums with bone rooms. Hrdlička responded simply to Heye, "As to our need of skeletal material, it is still a very great one."[90] Later, Hrdlička assured Heye, "though the specimens may be few in number, they will be of value."[91]

In the early 1920s, John P. Harrington, a linguist affiliated with the Bureau of American Ethnology, also briefly conducted archaeological fieldwork in California. Harrington was as much renowned for his brilliance in linguistics as he was for his hoarding linguistic data in his solitary office. Piles of notes were left untouched following extensive visits to the field. Slow to publish his discoveries in his own field, Harrington was notorious for territorially guarding the topics in his ongoing research in the anthropological community. Harrington, like so many other anthropologists of the era, occasionally collected skeletal material while in the field examining local languages and culture. He also occasionally sent boxes of bones to the museums.

Following the discovery of a pair of skulls at a site near Santa Barbara, Harrington submitted the remains to Kroeber at the University of California. Kroeber, trained broadly in linguistics, physical anthropology, and cultural anthropology, periodically assisted scholars like Harrington in the identification of materials, including skeletal remains, from the region of his expertise—California. Kroeber explained to Harrington that the remains were potentially "proto-Chumash"—an ancestor of a modern tribe of California Indians—and that the combined evidence of the associated artifact with the shape and size of the skulls confirmed this suspicion. Kroeber wrote, "Comparison of the measurements of your skulls with the measured Chumash skulls confirms my first impression that your two individuals fall within the limits of the Chumash type."[92]

Harrington, like Densmore and Heye, was not trained extensively in archaeology or physical anthropology. Nevertheless, such individuals within the larger anthropological community knew that human remains, especially remains of indigenous peoples, possessed especially valuable scientific potential. In each example, those who discovered the remains turned to another scholar more experienced in studying human remains—a scholar who happened to be affiliated with an expanding repository for remains in the form of a bone room at a museum. These three examples also point to the overriding desire to determine the exact race or tribal group represented by each set of remains and an effort to preserve remains for future generations of scientists in the face of what were believed to be rapidly disappearing archaeological and ethnographic sources. Despite the shared goals of preservation and racial definition shared by the nonarchaeologists collecting human remains, these goals did not necessarily align with the desires of physical anthropologists, who were intent on constructing overarching theories of race and human history as they raced to gather remains for their bone rooms. These three examples all took place as theories of racial classification were on the ebb in the United States, and they occurred as the physical anthropology community in the United States was finding itself increasingly concerned with questions about human history—such as the peopling of the Americas. News stories on Ishi and other purported discoveries may have further underscored the idea that scientists from museums were interested in studying the human body, but scientists and amateurs in the field were already well aware of bone rooms, despite an overall lack of *display* of bodies during this period.

In the correspondence related to the skeletal remains that Heye later turned over to the Smithsonian, he thanks Hrdlička for giving the bones "proper care and accommodation."[93] The language of salvage, preservation, and appropriate care for human remains was shared by anthropologists and museum leaders working in a variety of fields, indicating a common professional conceptualization of the rationale for acquiring and maintaining collections. Human remains, specifically those believed to be American Indian, were considered best collected from their place of burial and preserved at a museum for research and possible future display. The shared function of teaching scientists and broader public, while preserving specimens for future generations, was compelling enough to drive the construction of bone empires through the uncertainty of their exact utility for studies on race or human history. The manner in which this professional relationship extended beyond those who were primarily concerned with physical anthropology demonstrates the pervasiveness of this idea within the broader anthropological community. This broadly accepted paradigm lasted well into the twentieth century, and collections for select museums—including both the Smithsonian and the University of California, Berkeley—grew with particular rapidity.

DETERMINING THE VALUE OF REMAINS

One challenge facing museum professionals in the early twentieth century was how to determine the value of objects when negotiating exchanges. In physical anthropology, specifically, museums periodically exchanged specimens with other institutions on the basis of perceived redundancies or strengths within collections. Numerous examples of skulls, while valuable for comparative studies championed by racial scientists, were less valuable in showing the range of human diversity through exhibition, where typically only one or two examples from each group was needed to achieve the desired effect. Discussing a possible exchange with a museum in South America, Hrdlička explained, "A fair rate of exchange would be a skull for a skull, where crania alone are concerned, and two to three American skulls for each Kaffir skeleton according to its completeness and condition."[94] When a similar opportunity arose to obtain skeletal material from a museum in Siberia, Hrdlička proposed that the Smithsonian's Peruvian crania, "or first class busts of our aborigines," be exchanged for "well identified crania and bones of the Siberian natives."[95] It was unusual but also a genuine intellectual activity to judge the value of one set of bones against another. A similar opportunity for exchange arose in 1920, when Hrdlička visited Japan. After negotiating with Japanese officials, he proposed that twenty complete Japanese skeletons be exchanged for "a set of deformed crania and a few other things" that the Smithsonian could "easily spare."[96]

Museums determined the relative worth of human remains and material objects for value in exchange on the basis of three major factors: their rarity, their utility for science with regard to either racial classification or human history, and their worth for display in imagined future or current exhibitions. For Hrdlička, in the early years of the twentieth century, remains with some significance for the study of race were prized. Skeletons showing clear evidence of racial origin or those considered to be the purest racially—unfettered by interbreeding with other groups—were considered the most valuable for science. Gradually, his personal interest in the history of the peopling of the Americas assumed a greater share of his interest, and thus human remains that could provide evidence for the origins of American Indians became increasingly valued in exchanges. In the early decades of the twentieth century, with racial science at its apex in the United States and scientists increasingly turning to human remains to address questions about human history and evolution, bone rooms appeared to be assuming an increasingly central—if shifting—role in the future of American science.

In 1927, the physical anthropologist Henry Field began a letter to a fellow curator at the Field Museum of Natural History, "Just a note to say how pleased I am that you have 20 Eskimo skeletons! I hope you will bring back as many as you can because you know how empty those Cabinets are at the Eskimo end."[97] A common understanding of human remains as objects of scientific value had wholly permeated throughout the anthropological community. This common understanding that human remains might contribute to racial theory and knowledge of prehistory reached across disciplinary lines and allowed bone empires to grow with striking rapidity. The value of skeletons, though sometimes dictated by their utility for understanding such questions as the ancient arrival of people in the Americas, was most often calculated by their centrality in racial classification schemes. This reality, however, was about to undergo a major shift. In the early years of the twentieth century, museums, like the Field Museum and the Smithsonian, hoped to feature bone rooms that reflected the entirety of racial diversity in their specimen drawers. Cabinets began not only to fill "at the Eskimo end" but also to gradually start representing evidence from throughout time.

EUGENICS, OLD AMERICANS, AND BRAINS UNDER GLASS

The Smithsonian also entered the fray in the discourse surrounding eugenics.[98] Unlike other predominant theories in physical anthropology, eugenic ideas were displayed prominently at museums and fairs, in large part due to the efforts of eugenicists to advocate such theories. These displays sometimes even included notable, if modest, displays of human remains pulled from bone rooms. In 1912, with the looming arrival of the Congress of Hygiene and Demography, the Smithsonian planned a temporary exhibit for the visiting congress. In describing the scope of the planned exhibit, Hrdlička repeated his overall interest in the subject of eugenics and racial mixing, stating his wish to compare Americans of three or more generations—referred to as "thoroughbred American[s]"—to the offspring of "regular intermarriages between the Indians and the whites."[99] In addition to the intellectual connections drawn by curatorial staff between the study of race and the field of eugenics through text on exhibit panels, the Smithsonian also seemed eager to display a small portion of its growing collection of human bodies. Up to that time, these mummies and skeletons had been awarded little exhibition space at the national museum.

For a time, the studies of racial classification, prehistory, human evolution, and what was termed "hygiene and demography" shared the stage for those in anthropology who were interested in conducting research on the human body. A mere two weeks before the opening of the series of interdepartmental exhibits on eugenics at the Smithsonian, Hrdlička and the scholars Charles Peabody and George Grant MacCurdy traveled to Geneva for the Fourteenth International Congress of Prehistoric Anthropology and Archaeology to represent both the Smithsonian and the United States.[100] At the Smithsonian, the new exhibit examined race as its central theme, in large part because eugenicist organizations were pushing for these types of display. Smithsonian scientists and scholars were, of course, sympathetic to desires to utilize the expanding collections of human remains in this vein; however, they were also recognizing potential uses for the bones as tools for other kinds of research—ideas that were less interesting to the eugenicists.

As the relationship between eugenics and physical anthropology unfolded in the United States, it briefly grew to be symbiotic. In 1926, Hrdlička was contacted by the American Eugenics Society (AES), which was keen to add to its list of individuals available to lecture to eager audiences.[101] Hrdlička responded by noting that he did not lecture on the subject of eugenics specifically, but his other lectures on the study of physical anthropology would be of great interest to the same groups.[102] Hrdlička, as the most prominent physical anthropologist in the United States, possessed a significant relationship to the field of eugenics through the International Congress of Eugenics (ICE) and the AES. He had also allowed John H. Kellogg, the founder of the Race Betterment Society, to join the board of the *American Journal of Physical Anthropology*. Kellogg, a wealthy industrialist, subsequently funded the early efforts of the journal.[103] The connections to eugenics ran deep, in part due to the fact that the organizations offered funding for exhibition and publication. Although many scientists, like Hrdlička, were skeptical of some of the claims and activities of the most prominent eugenicists in the United States and Europe, they were eager to display and write about what they believed were neglected collections of skeletons in their museums. If museums like the Smithsonian refused to display bones and mummified remains directly, scientists could perhaps find support from outside organizations that were eager to highlight racial difference through science.

By late 1921, Hrdlička was working closely with the International Congress of Eugenics to organize displays at the American Museum of Natural History (AMNH) in New York. The Congress met at the AMNH, hosting displays in addition to the presentation of dozens of scientific papers from scholars from both the United States and Europe.[104] At the opening ceremonies for the congress, Henry Fairfield Osborn, a curator and prolific author from the AMNH, proclaimed, "I doubt if there has ever been a moment in the world's history when an international conference on race character and betterment has been more important than the present."[105] Osborn argued that the papers and exhibitions would demonstrate the stability of the races of humankind, with all of their "vices" and "virtues."[106] Despite signs of an apparent shift toward research centered on evolution and prehistory, race was still a hot commodity.

The exhibits displayed at the Smithsonian in time for the congress were largely based on Hrdlička's research comparing the physical features of recent immigrants and Europeans to the features of third- and fourth-generation citizens of the United States. He ultimately published his work in *The Old Americans*, appearing in 1925. The book argued that little change occurred in the physical structure of Americans over the course of a few generations, but it maintained that certain physical characteristics of North Americans would continue to differentiate from those of Europeans over time. Further, the book posited that those who had made it through the hardship of the immigration process were most likely "rather above than below the average in sturdiness and energy."[107] Hrdlička noted that many people had come to the belief that a distinctive type of American southerner, American westerner, and American youth existed, with people in various geographical areas advancing the belief that their own regional environment and generational culture crafted special forms of American robustness. He continued, "Suggestions have even been advanced that the American type is approaching that of the American Indian; the idea being presumably that since American environment produced the Indian—which in reality it has not done or not fully—it would in due time shape other peoples here to similar mold."[108] Certainly, many of these ideas had existed in some form for centuries, but much of Hrdlička's argument was directed against a claim made by Franz Boas, who studied recent immigrants and argued that physical features changed based on environment with striking rapidity.[109] Scholarly differences, in addition to what might be described as mutual stubbornness and irritability, led the two men to become increasingly distant and cold to each other. As theories of culture increasingly consumed Boas's attention and with his busy schedule of public speaking and training of students, Hrdlička became increasingly more isolated and firmly ensconced within the professionalizing subfield of physical anthropology. Anthropology of the era, in other words, continued to wrestle internally over the interpretations of anthropometric measurements and human remains. These debates were, at times, deeply influenced by prominent eugenicists, who funded exhibits and cited the findings of American anthropologists who generally supported their racialist ideas.

Hrdlička, basing his conclusions on the changing American body, detailed thorough measurements made between 1910 and 1924 of thousands of living humans and sets of human remains. He ultimately argued for the overall stability of racial characteristics while noting the changing nature of bodies in the United States over several generations.[110] The basic tenets of the ideas being put forward by Hrdlička and Boas regarding the body mirrored ideas inherent in the notion of American exceptionalism that was prevalent at the same time. Boas, however, leveraged his results to argue *against* the general stability of races, noting instead the influence of the American environment after just a few short generations. Nevertheless, the two were in agreement that *something* was taking place to change the nature of the body in the United States. At stake was the ability to place races into sturdy categories; if human races were constantly shifting and changing, scholars wondered if it would be possible to understand their "original" form after several generations of intermixing. These ideas of racial mixing and changing pressed scholars to collect as many "original," or "racially pure," skeletons as possible—prized both for their age and due to the belief that racial mixing had contaminated older physical forms, thus clouding what had been clearer racial lines. These same ideas also strengthened intellectual ties with the eugenics movement.

The exhibition resulting from Hrdlička's work on "the Old Americans" at the AMNH filled an alcove in (somewhat ironically) Darwin Hall at the museum.[111] Through the alcove, visitors walked past seven cases of displays. The displays began with a series of brain casts. The brain of a gibbon, an orangutan, a chimpanzee, and a gorilla were compared to the brains of several humans from differing races. The brains were intended to show "extremes of variation under normal conditions in brain evolution."[112] Following the display of brains, the variations of specific skeletal parts were compared in a series of five cases. Ribs, sternums, and femurs were lined up alongside numerous other bones, again intending to show normal human variation. From there, visitors would engage with a series of cases demonstrating the concept of reversion, or the return of a specific characteristic to a previously possessed ancestral form. One collection, in particular, featured a collection of American Indian skulls under the flat glass cases. Visitors, viewing the specimens from above, were instructed to observe in the skulls of Native Americans "the persistence to this day of Neanderthaloid forms and other primitive features."[113] Ideas about evolution, presented in this context, largely buttressed the idea that human races arose from earlier ancestors and might be understood, at least in part, from the observations of more or less "advanced" characteristics.

Importantly, the exhibition closed with a series of displays on the subject of heredity. To demonstrate this concept to eugenicists, Hrdlička designed a case featuring a collection of skulls from pre-Columbian Peruvian Indians. The collection originated from a single locality, and the crania were all absent an auditory apparatus on the right side of their skulls. This was a prehistoric example, but the remains were displayed in the context of a genetic lesson on heredity. The final two cases focused directly on Hrdlička's measurements of so-called Old Americans. They featured a comparative study of hair color and "a large chart showing the results of measurements and tests."[114] Most eugenicists of the era would have found the information presented in the displays compelling but perhaps not entirely satisfying. Though the displays pointed directly to the notions of American Indian primitivism and white superiority, the organizers appear to have hesitated in crafting any sort of stark claims about the inherent ability of particular races. Rather, claims for white supremacy were supported only indirectly and masked with a language of scientific authority and certainty.

A few years after designing displays for the American Museum of Natural History and the International Congress of Eugenics, Hrdlička articulated the belief that "eugenics in the future will be one of the fundamental subjects in all schools."[115] Hrdlička was certainly not alone in his interest in eugenics. In 1926, the American Eugenics Society listed as members dozens of preeminent individuals including professors, university presidents, clergy and medical doctors.[116] This momentary visibility for eugenics, and its connection to the practice of collecting and displaying human remains, was in reality short-lived. In the decades that followed, a growing interest in human evolution and prehistory took this practice in a new direction, but a racialist interest in science and classification theories of humankind momentarily fueled research and human remains exhibits in various settings.

. . .

By the 1910s and 1920s, important scholars in the United States, including Franz Boas and Alfred Kroeber, were already disputing the validity of race as a legitimate concept. In the context of the Americas, race seemed an unstable reality, quickly complicated by constant intermixing. These critiques stemmed from both changing social theories and scientific ideas that clashed with existing racial tenets of classification. Rather than fully engaging with these debates, many scholars argued that anthropologists should instead concern themselves with the idea of culture, as opposed to the concept of race. While the number of anthropologists who fell into the mold of Boasian anthropology grew throughout the first half of the twentieth century, many scholars continued to collect remains under the broad assumption that their collection and salvage would somehow be important for the study of race. Indeed, as the rate of procurement and measurement of collections increased, many seemed to have believed that the potential for unlocking secrets about race was much greater than ever actually achieved. Most skeletons stayed locked away in bone rooms, behind the scenes at a few major museums, subjected to few occasional displays and rare bouts of observation and study.

Despite growing bone rooms during the early portion of the twentieth century, museum professionals seemingly slowed their push to place remains on exhibit during this period.[117] Collections grew in places like Washington, DC, Chicago, and San Francisco, but bones were mostly held in storage, accessible to students and researchers but unavailable and virtually unknown to the typical museumgoer. Museum leaders generally abided by the grandiose visions of physical anthropologists only when outside funding, sometimes from eugenicist organizations, became available. Scientists proclaimed a steady interest in human evolution, and the North American continent was believed to be a more productive test ground for the subject of race, a concept dominating the scholarship in physical anthropology throughout much of the early portion of the twentieth century. Scholars, scientists, and amateurs traveling around the world were keenly aware of museums as sites for preserving certain kinds of skeletons—and that the bodies of indigenous or nonwhite peoples were considered especially useful as scientific tools.

Periodically, the popular media chronicled and celebrated discoveries of human remains. These accounts were often dramatized to the point of fiction, prompting one anthropologist to describe a newspaper article that described the discovery of several prehistoric skulls in California as containing "statements . . . about as idiotic as newspaper accounts usually are." The same anthropologist even continued by arguing, "Indeed, I think that such newspaper stuff does a good deal more harm than good."[118] Nevertheless, newspaper accounts of dramatic displays of recently discovered or acquired human remains brought visitors to museums and fairs in droves, and they lent legitimacy and attention to those scholars concerned with collecting remains. New exhibits were modest in number, but the cresting interest in the project helped museums develop a foundation for later public display.

Although the practice of collecting and measuring human bodies was never a driving force behind the work of people like Densmore, Fletcher, or Harrington, all three did collect human remains on behalf of museums. Museums responded by readily adding the skeletons to their collections. The sharp division that historians have drawn between the heroic and sympathetic ethnologist, fighting for Indian rights, and the evil "bone collector" proves to be an inaccurate division. The feeling of urgency to collect human remains was felt by a wide range of scholars, amateur collectors, and donors all around the United States. Importantly, these individuals brought the practice of collecting bodies around the globe and forced new legislation. Despite early success in establishing bone collections from increasingly distant places, museums emerged as spaces for expressing changing ideas about the human body in certain spaces. When these bodies were actually placed on exhibit, however, they became powerful and fraught symbols that were indented to stand in place for the ideas of those who displayed them.

CHAPTER 3

The MEDICAL BODY

on DISPLAY

As new ideas about race and the human body emerged in the nineteenth century, several small medical museums opened in the United States, mimicking older institutions existing in Europe.[1] Often affiliated with hospitals or medical schools, medical museums became critical tools for teaching students about various conditions that were not typically seen during hospital rounds. Medical collections transformed body parts into *specimens* used for teaching, presenting key lessons about the human body to future physicians.[2] Bodies and body parts, showing common or rare pathologies, were preserved with varied success through different drying methods and chemistry. The techniques attempted to stave off the natural decay of the human corpse.[3] Preserved specimens, in addition to cast and model collections, could be gingerly passed around a classroom or placed under glass and easily exhibited.[4] Unlike illustrations or photographs printed in textbooks, medical specimens could be experienced in three dimensions, allowing students to better comprehend the condition. Dissections, important in medical education, were messy and cumbersome procedures, especially when conducted by clumsy and inexperienced students.

Further complicating matters was the fact that examples of particular pathologies could not always be repeatedly secured for each new class of medical students to dissect—another frequent problem in medical education during the nineteenth century. Medical specimens, on the other hand, could be isolated, cleaned, and preserved for future students or the layperson in exhibitions. Collecting and preserving human remains for permanent storage and display appeared to be the next logical step in modernizing medicine in the United States and Europe. Yet for all the potential of medical museums, their leaders in North America complained, "The important functions of the Medical Museum as a compendium of scientific facts, a storehouse of material for research-work, and as a teaching medium are, on this continent especially, comparatively little appreciated."[5]

Despite rising support for medical museums, the medical community in the United States struggled to articulate an exact role for these institutions within medical education and practice—questions about appropriate audience loomed especially large over these museums as they grew. Museums collected and displayed wide-ranging natural human variation, as well as highly unusual examples of bodily manipulation and disease. Abnormal medical specimens, models, and human remains also appealed to an interested public. Eventually, a mission geared more toward public health replaced the role these museums assumed in the education of medical professionals.[6] Medical museums helped satisfy visceral desires and macabre curiosities. As they became increasingly open to the public, medical displays subverted dominant social norms by placing bodies, body parts, or visible pathologies considered too shocking for public display into a more socially acceptable context. Even as they became more open to broader audiences, medical museums continued to be reflections of the medical gaze, constructed and organized by curators who frequently benefited from collections acquired by individual physicians with little thought as to obtaining permission from the dying or their next of kin.[7] While race and history proved less central a focus in medical museums than in settings more related to natural history—where bodies were rarified as evidence in support of classificatory theories in subtle yet important ways—remains were frequently catalogued and exhibited with specific reference to race. Over time, racial classification schemes offered by anthropologists permeated medical museum collecting practices. Even in vastly different cases, the end result proved similar: skulls, bones, and to a lesser extent mummies steadily came to museums, where they were catalogued, stored, studied, and delicately placed on exhibit.

Medical displays and exhibits appeared in a variety of forms and sizes in the United States throughout the nineteenth and twentieth centuries. Ironically, just as medical museums declined in number, they became gradually more accessible to public audiences. One of the largest and best known medical museums in the United States is the Mütter Museum in Philadelphia, Pennsylvania. Though specimens that were determined to be racially or historically significant were often marginalized in medical museum collecting, over time, medical museums such as the Mütter Museum *did* occasionally acquire such specimens. The act of acquiring and organizing certain bodies points to an opportunistic medical collecting through the growth of a privileged scientific elite. But collecting was frequently outpaced by shifting theories in medicine, which experienced a series of radical transformations between the eighteenth, nineteenth, and twentieth centuries.[8] Acquiring, classifying, and exhibiting human remains under the guise of medical specimens contributed to the construction of cultural ideas related to human body and racial difference.

As seen earlier, the Army Medical Museum (AMM) focused its early collecting on pathological conditions created by battlefield conditions and comparative racial studies. Medical museums like the AMM waxed and waned in their emphasis on acquiring skeletons to explore the races of humankind. This was based not only on the opportunistic museum collecting habits of the era but also on the interests of the museum leaders. At the turn of the century, museums were still sorting out where each skeleton might best be stored for science. In 1897, when the AMM divested itself of all nonpathological remains, the institution receiving them, the Smithsonian, started a new physical anthropology division for the primary purpose of addressing race in comparative terms. The AMM, on the other hand, shifted its emphasis to samples of pathologies treatable by modern medicine. But medical museums did not totally divest themselves of the race question. Not only did medical museums seek to define the "normal" and "abnormal" condition of the human body; they also occasionally sought to define patterns of difference between perceived races.

MEDICAL MUSEUM FOUNDATIONS

The College of Physicians of Philadelphia emerged as a private medical society in 1787. Years later, in the 1840s, the college established a permanent pathology collection.[9] Shortly after establishing the collections, fellows of the college came forward to donate their own private materials, acquired during their own respective tenures as physicians. Doctors donated, in the words of one historian, "gallstones, monsters, and plaster casts of one condition or another."[10] Despite the early success that the museum demonstrated in acquiring medical instruments and specimens, interest in the collection among the fellows rapidly declined. In 1858, a professor from Jefferson Medical College, Thomas Dent Mütter (1811–1859), reinvigorated the collection by donating 1,344 objects and contributing $30,000 toward an endowment. Mütter was ill, and he wanted his personal collection of medical photographs and specimens to be available to medical students and physicians for professional education.[11] The donated collection included bones, wet preparations, and casts, as well as a number of paintings.[12] Early growth of the collection occurred just as anatomists were improving their ability to preserve the fleshy parts of the body, rather than just dried specimens and skeletons.[13]

The Mütter Museum opened a new building in 1863, just one year after the Army Medical Museum opened. Initially, the public was not allowed to view the growing pathological collection in Philadelphia. The small museum was reserved only for fellows of the college. In 1867, the Anatomy Act of Pennsylvania allowed medical schools and societies in the Philadelphia region (including the College of Physicians of Philadelphia, which governed the Mütter Museum) easier, and legal, access to human cadavers for dissection and even permanent preservation as medical specimens.[14]

After the Civil War, medical museums gradually moved from specialist institutions reserved for the medical elite to sites intended for the diffusion of knowledge regarding public health. By 1883, the president of the College of Physicians of Philadelphia, Alfred Stillé, argued, "The day is passing when it is any longer necessary to hold a dark screen between physicians and educated laymen, and the better the world shall learn what are the aims and the achievements of a physician's life, the sooner will be dispelled the prejudices that antagonize the medical profession and the delusions under which the public became the victims of error and of fraud."[15] Just as physicians in the U.S. military were asked to contribute skeletal material to the AMM, physicians affiliated with the College of Physicians of Philadelphia came forward with medical specimens for the collection.

Over ensuing decades, the Mütter Museum, imitating medical museums in Europe, collected and exhibited unusual pathological specimens. Like the Army Medical Museum, the Mütter Museum collected a number of sizable collections of crania; however, the main focus of the early displays was to educate medical students and physicians about human diseases rarely seen in most medical offices.[16] Medical students might learn about the effects of a disease by passing around and carefully examining a human skull of an individual afflicted with the ailment during life. Over the course of the first few decades of the Mütter Museum's existence, however, its audiences were still fairly limited, with collections only periodically shown to fellows or small groups of medical students.

Similarly, the Army Medical Museum's earliest collections mostly languished in storage. When the AMM moved to Ford's Theater in 1866, the utility of the collections became clear almost immediately when museum officials began experimenting with regular exhibitions. The surgeon general stated that civilian and military medical professionals consulted the collections "weekly and almost daily."[17] Army officials found the use of the collections by medical professionals to be encouraging—especially in light of the fact that efforts to create the Army Medical School were stymied throughout most of the second half of the nineteenth century.[18] Having opened in the midst of the Civil War, the Army Medical Museum was also consulted by unique visitors: soldiers who had lost their own limbs during the war. A veteran named J. F. Allen, for instance, recounted his story of finding his own arm, amputated during the Civil War, preserved in the medical museum.[19] Journalists were generally amazed not only by the display of the bodies of the dead but also by the occasional connection of museum skeletons to the living. The documentation that the museum possessed detailing how a limb had been acquired—an effort to understand the medical history of the individual—could actually help veterans, in certain instances, prove to government agencies that they had been active in their service during the Civil War and should therefore qualify for benefits. One physician described the individuals who came to view their lost body parts as "officers and soldiers who had lost a limb by amputation," who would come to the museum "to look up its resting place, in some sense its last resting place."[20]

Narratives of soldiers who lost limbs and their ties to medical museums even assumed a place in fiction. In 1866, S. Weir Mitchell, a physician, prominent fellow of the College of Physicians of Philadelphia, and prolific author, wrote a short story about a young army surgeon named George Dedlow. Mitchell's stature in the medical community was strong following the Civil War, and he is best remembered as the creator of the "rest cure" for female patients suffering from neurasthenia.[21] He also dabbled in fiction. His works include "The Case of George Dedlow," a short story elegantly examining a soldier's loss of identity following the removal of his limbs during the war; the story ties this displacement of identity to the rise of the medical museum.[22] Dedlow joins the infantry during the Civil War and loses all four limbs to painful amputation. The graphic story goes into great detail, explaining how each limb was harmed through enemy action or infection, the protagonist enduring the grueling removal of his limbs without anesthesia. The fictional account of surgical amputation from the perspective of the patient is the first recorded reference to the phenomenon of *phantom limb*—an occurrence following amputation in which the patient continues to feel sensation in the lost body part. In the story, Dedlow is eventually transferred to a hospital in Philadelphia intended to aid soldiers who lost limbs during the war—based on an actual clinic that Mitchell helped organize during the conflict—and the account reaches its dramatic climax when the protagonist is asked to take part in a séance. Around a circle of soldiers participating in the séance, the spiritual medium receives a message of a pair of numbers. Spirits communicate through the medium a series of museum catalogue numbers, spoken before the group: "UNITED STATES ARMY MEDICAL MUSEUM, Nos. 3486, 3487." Dedlow exclaims in response, "Good gracious . . . They are *my legs—my legs!*" Dedlow then recounts, to the astonishment of everyone in the room, that he was briefly able to walk across the room using his invisible legs, which the medium had summoned from the museum in spirit. Gradually, however, the power of the invisible, reanimated legs began to fade, and Dedlow slowly sinks to the floor—once again a shadow of a man without the use of his limbs.[23]

The Army Medical Museum, around the time of the publication of Mitchell's story, was quickly starting to attract tourists from the broader American public, beyond physicians and medical students. Despite the reservations of tourist guidebooks and the museum's several moves to new locations and limited times open to the public, the AMM drew a surprisingly large number of curious visitors.[24] The desire to view lost limbs, deformed abnormalities, and the skeletons of races from around the world all proved to be captivating draws for popular audiences. Journalistic accounts of the museum in this period point to the possible metaphysical meditations on life and death that a visitor might engage in while walking through the modestly sized galleries. Physicians, on the other hand, worried that visitors would be drawn to medical museums hoping only to view macabre abnormalities or grotesque deformations. Whatever the motivation for entering the galleries, visitors from off the street soon easily outnumbered professional visitors who hoped to consult or research the collections. Reading about the Army Medical Museum in captivating stories like "The Case of George Dedlow," the public was encouraged to tie together an appeal for the bizarre with a curiosity regarding modernizing medical science.

Medical museum exhibitions expanded on the much older practice of medical illustration. Pictures were easier to disseminate, but physical examples of specimens were still viewed as a valuable complement to medical instruction. Medical photographs and exhibitions of actual specimens, unlike illustrations and paintings, came to be considered more objective conveyers of truth. Similarly, wax models were considered to be stable, consistent alternatives to actual human remains. While the medical illustration or photograph could capture the essence of a particular medical condition, medical museums were founded on the assumption that physical records of medical conditions were more powerful and accurate pedagogical tools. Medical tools, wax models, and illustrations might add significantly to collections, but specimens of human flesh and bone were the most critical aspect of the medical museum collection.[25] Medical museums might have been mostly closed off the public, but special exhibits again proved the consistent existence of public curiosity.

MEDICAL DISPLAYS AND WORLD'S FAIRS

Between 1876 and 1904, the United States hosted a series of major international expositions in which human remains prominently displayed advances in medical science. While many of the skeletons displayed at world's fairs were displayed under the guise of the emerging field of anthropology, the changing nature of medicine in the United States, spanning the Gilded Age to the Progressive era, encouraged additional displays on the subject of medical health. New discoveries in anesthesia, prosthetics, and public health were touted, and public medical exhibitions stood near the displays created by medical corporations, many of them unveiling new products to fairgoers. The historian Julie K. Brown traced the development of these displays, arguing that they contributed to a broader cultural conception of individual and national health emerging in the United States, especially between 1876 and 1904. In addition to presenting novel ideas about health and hygiene, exposition organizers highlighted new urban sanitation systems.[26] The era was also marked by a significant growth in eugenics. At the same moment, a broadly conceived notion of hygiene and public health emerged in major urban centers, encouraged in the United States by both public and private organizations that learned from their counterparts in Europe.[27] All of these ideas came to bear both on the global stage of international expositions and in public health campaigns and exhibits of a wide variety and size.[28]

In 1876, at the Centennial Exposition in Philadelphia, an enormous twenty-one-acre facility—the main building of the exposition—was dedicated to the topic of health. Within this large building, visitors found fifty-two individual exhibitors displaying commercially available medical products, prosthetic limbs, and surgical devices.[29] One display, organized by Adam Politzer, a professor of otology at the University of Vienna, caught the particular attention of the curators of the nearby Mütter Museum. Politzer had created successful displays for the 1867 exposition in Paris; in Philadelphia, he designed a display of dissections "illustrating normal and pathological anatomy of the human ear."[30] A curator from the Mütter Museum purchased the collection for the considerable sum of $800.[31] The fine print of the transaction, however, designated that the cost of the collection was not for the specimens of human ears itself, which Politzer had acquired from his own patients, but rather was a fee for the custom mounts of the specimens.[32] This particular detail of the acquisition points to lingering questions surrounding the ethics of acquiring human remains in the wake of changing anatomy laws.

The World's Columbian Exposition held in Chicago in 1893 featured an even larger and more diverse series of medical displays. As described earlier in this book, the massive fair featured numerous displays of human remains in various contexts. Recently discovered mummies from the American Southwest attracted the attention of an audience that had read extensively about the rediscovery of cliff dwellings but had yet to view them firsthand. Displays of recent finds from the American Southwest were both public and private, for-profit enterprises—creating a striking mixture of display methods that seemingly borrowed at times from the circus-like atmosphere of P. T. Barnum's museum and the rapidly professionalizing nature of nineteenth-century museum anthropology. The anthropology building at the fair featured small displays on physical anthropology, mainly on the heavily racialized practice of anthropometry, or the study of the shape of the bodies of the living. Adding to the mixture of anthropological, or pseudoanthropological, displays of human remains were separate exhibitors who hoped to display *medical* ideas at the World's Columbian Exposition utilizing human remains as well. Medical products, which had been placed in the manufacturing section of the 1876 Centennial Exposition, were moved to the Liberal Arts Department in Chicago. While many of the displays were intended to function as advertisements aimed at medical professionals visiting the fairgrounds, other displays were geared toward the lay visitor.[33] The Army Medical Department sponsored a small exhibit featuring eighty bone sections that demonstrated the effect of bullet-wound ballistics on the human body. As noted earlier, the AMM had started to shift its focus away from collecting large numbers of skeletal remains toward soft-tissue samples that catalogued evidence of communicable diseases. In the museum's small exhibit case at the World's Columbian Exposition, examples of bullet wounds on human bones were both prominent and striking, but images of bacteria and other soft-tissue samples lined the walls behind the case. Notably missing from the AMM displays were studies centering on "comparative anatomy"—or the comparative study of race—which was awarded a prominent place in the early collections of the museum.[34]

At the Louisiana Purchase Exposition in 1904, as Aleš Hrdlička of the Smithsonian busied himself measuring living visitors and collecting the remains of unfortunate indigenous peoples who died at the fair, exhibitors were displaying a number of pathology and anatomy galleries. State health departments were encouraged to contribute exhibitions to the fairgrounds, many of which focused on the management of public health crises, one example being tuberculosis. (Anthropologists, too, were concerned with tuberculosis, yet they framed the disease in terms of the supposed decline of the American Indian.[35] In the medical displays at the 1904 fair, the disease was presented with little or no emphasis on race.) The physician and anatomy displays also featured materials brought together from a number of hospitals, laboratories, and private collections. Just as in 1876 and 1893, collections from the Army Medical Museum were again featured prominently in the 1904 fair. These displays included, in the words of a recent historian of the medical displays at the fair, "dissections of all parts of the human body, parallel dissections of the lower animals, and beautifully mounted pathological specimens of diseased conditions of the human body."[36] An estimated 800 visitors per day, or about 144,900 total visitors during the run of the 1904 St. Louis fair, toured the medical displays there; it was assumed that many were either physicians or dentists.[37]

Medical exhibits at international expositions held between 1876 and 1904 grew in both size and sophistication. While certain exhibits did indeed hint at ideas about race, gender, or prehistory in small ways, these themes were noticeably absent in most medical exhibits. Other exhibitions—specifically those of anthropology or others more focused on eugenic ideas under the label of "hygiene"—pointed more directly to ideas related to racial classification. Displays that focused on medicine at international expositions had two important features in common with natural history museums and permanent medical museums. First, the displays at world's fairs often featured material loaned to the fairgrounds from permanent museums. Additionally, while these fairs did embrace the use of models and casts, nothing was as compelling or captivating to professional audiences as actual human remains.

Clever, mild mannered, and affable, William Henry Holmes (1846–1933) made a career as a museum builder. Hired by the Bureau of American Ethnology, he ultimately became a leader in the Smithsonian Institution and arranged for the transfer of the human remains collections from the Army Medical Museum to the Smithsonian. Photograph of miniature on paper by Alyn Williams, undated. National Anthropological Archives, Smithsonian Institution, [NAA INV 02859100, Photo Lot 33].

Hired by the Smithsonian in 1904, the Czech-born Aleš Hrdlicka (1869–1943) transformed the Smithsonian's physical anthropology collections and sparked a global competition over human remains. Intensely passionate, smart, and sometimes deeply divisive, he had an influence that was significant despite the fact that many people in the scientific community came to reject his ideas. Smithsonian Institution Archives, Image #SIA2009-4246.

Newspapers fueled the movement to collect human remains. They largely depicted the skeletons as rare and valuable commodities for emergent sciences. Notably, newspapers frequently framed stories about discoveries around the supposed racial origins of human remains. This Los Angeles newspaper features a large cover story about the cliff dwellers of the American Southwest. *Los Angeles Herald*, July 2, 1905, Sunday supplement.

A physical anthropology storage room in 1911. Opening a new building in 1910, the Smithsonian's National Museum of Natural History had space to sufficiently organize its human skeletal collections. Soon, however, so many bones were being sent to the museum that storage again became an issue. How the Smithsonian organized, catalogued, and stored human remains became a model for other museums across the United States and around the world. Smithsonian Institution Archives, Image #24048 or NHB-24048.

The Piegan Blackfoot Mountain chief listens to a recording made with the ethnologist Frances Densmore (1867–1957). Densmore was one of the most important ethnomusicologists of her era, but she was routinely denied resources to complete her work by government agencies in Washington, DC. Lesser known is the fact that Densmore collected human remains during her fieldwork and shipped them to the Smithsonian, where they were added to the permanent collections. Library of Congress Prints and Photographs Division, LC-USZ62-107289.

Henry B. Collins, Jr. (1899–1987), with a skull collection. In the decades following the initial organization of the physical anthropology department at the Smithsonian, collections grew with unparalleled momentum. Museum leaders encouraged archaeologists like Collins to collect human remains in addition to other material culture discoveries made during fieldwork. Smithsonian Institution Archives Image #SIA2009-2053 and 4892.

Alfred Kroeber became one of Franz Boas's most prominent students, focusing much of his attention on the history and culture of California Indians. Despite documenting his opposition in a letter from abroad, he ultimately facilitated the donation of a human brain—that of his friend Ishi, the last member of the Yahi and sometimes called the "last wild Indian." The reason behind his sudden change of heart on this matter has been a subject of debate. National Anthropological Archives, Smithsonian Institution, [NAA INV 02860700, Photo Lot 33].

An archaeologist and great-nephew of the department-store magnate Marshall Field, Henry Field (1902–1986) studied for a doctorate at Oxford University before returning to the United States to become the curator at the museum in Chicago bearing his family name: the Field Museum of Natural History. Field found himself both emulating and competing with the Smithsonian Institution with regard to human remains collections and new exhibitions. Courtesy of The Field Museum, copyright © The Field Museum, digital ID #GN78453.

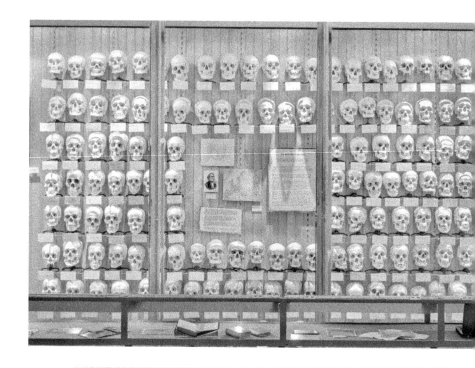

The Hyrtl Skull Collection at the Mütter Museum. Rarely are skull collections displayed today as they were in the nineteenth century. The Mütter Museum, however, displays its Hyrtl collection in an effort to teach not only basic human anatomy but also the history of medicine and scientific racism. Copyright © 2009 George Widman Photography, for the Mütter Museum of The College of Physicians of Philadelphia.

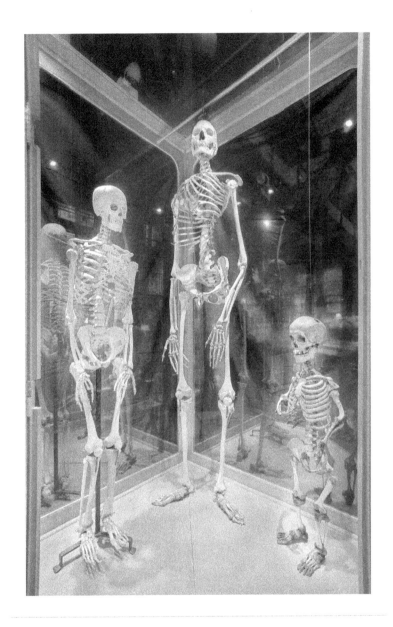

The skeletons of a giant, a dwarf, and an average human, Mütter Museum. Collected under mysterious circumstances, the Mütter American Giant offers a fascinating glimpse into the sometimes mysterious world of medical specimen collecting in the late nineteenth century. The man—believed to have died between the ages of twenty-two and twenty-four—stood well over seven feet tall during life. The purveyor offered the skeleton for sale, which the museum purchased for fifty dollars, under the condition that no questions were to be asked about the exact identity of the deceased.

Copyright © 2009 George Widman Photography, for the Mütter Museum of The College of Physicians of Philadelphia

The Story of Man through the Ages, opening in time for the 1915 Panama-California Exposition in San Diego, California, offered many Americans their first glimpse into human prehistory through artistic reconstructions based on early fossil finds. Copyright © San Diego Museum of Man.

Although the exhibit was later critiqued for including reconstructions based on fraudulent specimens, the reconstructions of human ancestors in San Diego were dramatic and captivating enough to hold attention. Several are still displayed at the San Diego Museum of Man. The busts are now described in their historical context, and the museum still considers them an important part of its original collection. Copyright © San Diego Museum of Man.

Thought to be the largest exhibition created on physical anthropology and human evolution to date, *The Story of Man through the Ages* prominently focused on the question of race. Several sections of the exhibition, however, introduced museum audiences to the story of human evolution and prehistory. Struggling to obtain permanent exhibit space at the Smithsonian, the special exhibit in San Diego allowed Aleš Hrdlicka and his collaborators to test certain ideas about display for the first time. Copyright © San Diego Museum of Man.

Casts of individuals at different ages were displayed prominently in *The Story of Man through the Ages* exhibition. Following the dominant thinking at the time, the races of North America were divided into European Americans, African Americans, and Native Americans. Despite noting the exhibit's flaws, Aleš Hrdlicka and his collaborators considered it a great success. Copyright © San Diego Museum of Man.

The final room in *The Story of Man through the Ages* was perhaps the most dramatic. Despite an uninspired presentation in flat exhibit cases, human remains from dozens of individuals captivated the large audiences that visited the fair. Copyright © San Diego Museum of Man.

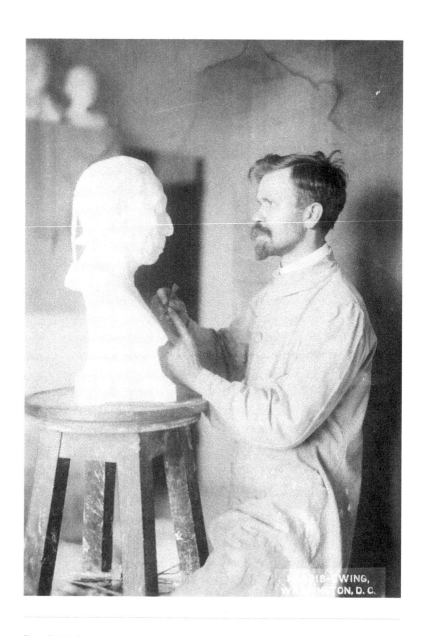

Frank Micka, who was hired in part because of his Czech background, ultimately modeled dozens of busts for *The Story of Man through the Ages* exhibition—later the San Diego Museum of Man. Henry Field, who visited the San Diego museum for ideas when he was planning major new displays for Chicago's Field Museum, criticized the Micka busts as lifeless, flat, and absent expression. National Anthropological Archives, Smithsonian Institution, [SPC010285.00 Neg 86-500].

While founding an important new training center for students interested in physical anthropology and comparative anatomy during a segregated era, W. Montague Cobb (1904–1990) became part of an important "Howard Circle" of leading African American intellectuals. Cobb used many of the same methods that had advanced the scientific racism of previous eras to break down the notion of racial difference.

Courtesy of the Moorland-Spigarn Research Center, Howard University Archives, Washington, DC.

William H. Egberts with skulls, 1926. As the physical anthropology collections grew at the Smithsonian Institution, researchers began using the bones to study questions other than race. Trepanning, the surgical technique used since ancient times, left distinct marks on the skulls of those who received the treatment. Library of Congress Prints and Photographs Division, LC-DIG-npcc-15637.

Henry B. Collins, Jr., in Alaska, 1929. Encouraged by Aleš Hrdlicka, Smithsonian archaeologists like Collins were keen to collect human remains during fieldwork. Here, Collins examines skulls before preparing them for shipment back to the Smithsonian. Smithsonian Institution Archives, Image #SIA2009-2048 and 4889-B.

As Aleš Hrdlicka grew older, his colleagues described him as increasingly demanding and difficult to work with. He continued to study the human remains collected for the Smithsonian Institution until his death in 1943. Smithsonian Institution Archives, Image #SIA2012-6453.

Mildred Trotter (1899–1991), who later became an internationally regarded expert in studying human blood and hair, encountered Henry Field at Oxford University, where she continued her training using a National Research Council Fellowship. She spent most of her career teaching at Washington University in St. Louis. Smithsonian Institution Archives, Image #SIA2010-0139.

DETERMINING THE CAPACITY OF HUMAN SKULLS IN PHYSICAL ANTHROPOLOGY

T. Dale Stewart tests the capacity of human skulls, Smithsonian. Aleš Hrdlicka was notoriously protective of the physical anthropology collections at the Smithsonian, but over time, Stewart earned his trust, eventually replacing his mentor as curator at the museum. Later, Stewart became the director of the National Museum of Natural History. Smithsonian Institution Archives, Image #2012-6451.

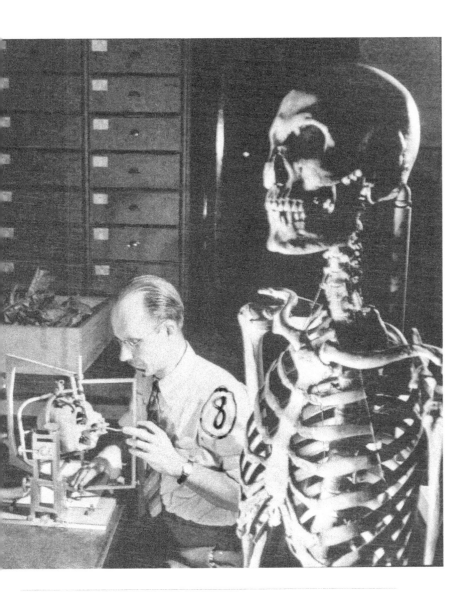

Despite having trained with Aleš Hrdlicka, T. Dale Stewart became decidedly less controversial than his mentor was. He was known to be detail oriented and meticulous in his studies. This staged publicity photograph was intended to give the public a glimpse into Stewart's day-to-day work with human remains at the museum. Smithsonian Institution Archives, Image #85-8067.

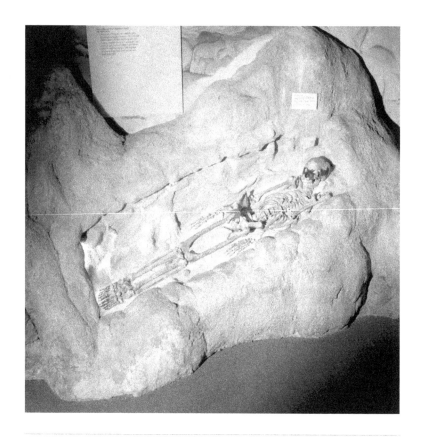

Magdalenian Girl, Field Museum. Some sources suggest that Magdalenian Girl was quietly shipped to the United States from Europe disguised in a flag-draped coffin, usually used for a fallen American soldier from the First World War. Following media reports that appeared in advance of her arrival, Chicagoans became fascinated with the skeleton. The skeleton was considered one of the most important "specimens" in any museum in the United States, and its display broke museum attendance records. Courtesy of Photographer John Weinstein. The Field Museum, copyright © The Field Museum, digital id #GN88907_2.

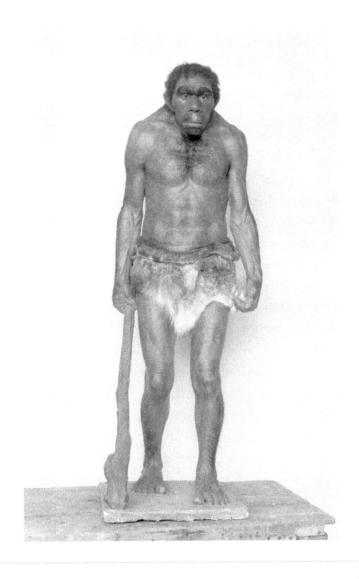

Front view of the figure of a Neanderthal man with a club, wearing animal-skin loincloth, for the Family of Gibraltar group, Field Museum. The Neanderthal figures modeled for the Field Museum's *Hall of Prehistoric Man* influenced how generations of Americans visualized human evolution. The dioramas remained on display for several decades. Diorama by Frederick Blaschke, Chicago, Illinois, 1932, Field Museum Library, Getty Images.

ACQUIRING BODIES FOR MEDICINE

In 1811, Chang and Eng Bunker were born conjoined at the torso. The pair originated in Siam (now Thailand), and the term *Siamese twins* was coined to describe them as they toured the globe as a curiosity—the term followed them their entire lives. Originally displayed as a minor sideshow oddity, the brothers eventually supported themselves through a successful enterprise exhibiting the unusual condition afflicting them. The two emigrated from Thailand to the United States in 1829 at the age of eighteen and died just a few hours apart in 1874.[38] Eng, whom the autopsy described as the "more excitable" of the twins, suffered from alcoholism, and the physicians were curious as to how his drinking would affect their conjoined bodies. During the course of the brothers' lives, men of medicine had also grown curious about the band of tissue connecting them—debating whether it would even be possible to successfully separate the brothers. When Chang died quietly in his sleep one evening, Eng's fate was sealed; he too soon died.

Once the pair died, some controversy ensued over the proper treatment of their bodies.[39] Fifteen days after their death, their bodies arrived at the Mütter Museum, where an autopsy was performed by a group of physicians as over one hundred fellows of the college looked on.[40] The *New York Times* covered the story of the brothers' arrival in Philadelphia on the front page. While the *Times* considered the story a significant enough subject for a front-page article, the condition of their bodies caused the paper to warn against any desire among readers to see them firsthand, stating, "The twins were not lovely in life, and in death their appearance is repulsive. Every ugly feature and uncultivated type seemed to have been made more strongly marked by the hand of death." The reporter added, "Thousands of people are burning with curiosity to view them. To all such I would say, 'Do not do so.

They are not desirable objects of reminiscence.'"[41] According to the autopsy report, no embalming procedure had been performed on their bodies.[42] In life, the twins had been portrayed as a mysterious and exotic medical mystery; in death, they were viewed as a repulsive yet valuable foreign specimen.

Over two days, as the twins' bodies decayed, several plaster casts of them were made, and the tissue connecting them in life was preserved in death.[43] The physicians who attended the autopsy finally had an opportunity to view each part of the unusual band connecting the brothers, as layers of blood and tissue were slowly wiped and peeled away.[44] Their bodies were also photographed and illustrated as the fellows examined them in detail. They had been viewed in life as something of a sideshow curiosity, and a portion of their mortal remains were placed on display in the Mütter Museum, utilized in teaching both professional and popular audiences about conjoined twins. The case of Chang and Eng, unlike the vast majority of specimens acquired by medical museums, was marked with celebrity status (it was rumored that police surrounded the building the night of their arrival in order to avoid foul play).[45] Despite the unusually high-profile nature of the case, the acquisition of the specimen was typical in that their bodies were used as the basis for casts or models following an autopsy, and a small portion of their remains were preserved and designated as a valuable "specimen."[46] The performative nature of their race that was presented during their life was partially displaced by medical curiosity upon their death. Demand for medical information was so great that the publication of their autopsy report caused some controversy between the college and several rival medical journals.[47] Writers and the audiences soon pouring into the museum to tour the displayed connective tissue and casts seemed drawn simultaneously to their condition and to their subtly racialized exotic otherness as recognizable showpieces at the museum. Even the conclusion of a medical dispatch describing Chang and Eng's autopsy sketched their medical condition into prescribed categories: "There are few double monstrosities so well developed as this one. I think the records of about four hundred monsters have now been collected, but very few are of such a complete nature as this. Every one has heard of the Hungarian Twins, who lived to the age of twenty-one years, in the last century."[48]

In the late nineteenth century, cities along the eastern seaboard passed a series of anatomy acts allowing for the legal acquisition of cadavers by specific medical institutions. Despite legal protections allowing certain medical schools the acquisition of remains, the demand for cadavers in 1879 already stood at over five thousand per year, and legal means of acquiring human bodies—mainly through almshouses and prisons—was outpaced by the demand for bodies, with body snatching continuing for a number of years.[49] On the other hand, medical specimens that had already been treated or preserved and that were intended for the teaching of medical students, also called *preparations*, could be purchased with relative ease. In 1877, after viewing the skeleton of a man estimated to have died between the ages of twenty-two and twenty-four, Mütter Museum curators decided to purchase the remains. The skeleton was of an exceptionally large individual, approximately seven feet six inches tall, and the remains were briefly on display at the Academy of Natural Sciences in Philadelphia before being transferred to the Mütter Museum at a cost of fifty dollars. A professional purveyor of anatomical specimens, who had entrepreneurially sold specimens to museums and schools, had prepared the skeleton by cleaning it and wiring it together. The purveyor offered the skeleton for sale, with the specific condition that no questions were to be asked regarding the exact identity of the deceased. Usually, when no medical history could be offered of specimens, medical museums found them to be less desirable. The highly unusual size of the giant, however, made the skeleton too hard to pass up. The so-called Mütter American Giant was of keen interest to the museum, which utilized the skeleton in creating displays about pathology that results in unusual size.[50] To an even greater extent than in the case of the "Siamese twins" Chang and Eng, the medical display of the giant skeleton deemphasized the man's ancestral background. While natural history museums might occasionally acquire unusually large or small human skeletons, for instance, their main goals were to collect racially or historically significant specimens—not medical anomalies. Certainly, pathologies were collected by anthropologists, but the acquisition and presentation of the Mütter American Giant is clearly unique to the medical museum.

Specimens came to the Mütter Museum through different means.[51] The most common was the outright donation of materials from private individuals—medical doctors turning over specimens they preserved themselves. The Mütter Museum made special effort to solicit donations from fellows of the college. Occasionally, physicians even prepared specimens with the museum in mind.[52] When physicians associated with the college traveled to Europe, they sometimes offered money from the museum to purchase particularly prized models and specimens.[53] Medical associations donated yet more specimens, books, and medical illustrations.[54] Although the original funds bequeathed by Thomas Dent Mütter were provided with the intention of constructing a new building for the museum, collections were purchased from the trust throughout the nineteenth century and into the twentieth.

At the turn of the century, the College of Physicians of Philadelphia made a point to remind its fellows that it was interested in obtaining new specimens for the collections. The museum successfully acquired a number of rare or special specimens, but the administrators of the museum were starting to come to the belief that individual specimens, "when classed with similar ones and studied in groups large enough [might] admit of comparison as to the points of variation from, or conformity to a prevailing type."[55] The philosophy driving acquisitions—the belief that greater numbers of specimens could help in the systematic classification of different types of pathology—echoed the taxonomic drive among the scholars who supervised collections at natural history museums. The racially specific language that worked to define particular bodies made its way into medical museum practice both haphazardly and systematically. An early museum catalogue listing skeletons acquired for the Mütter Museum, for example, lists "fragments of Indian bones found in a mound near Luray, VA," among a list of fragmentary skeletons with no other references to race in the records. While the racial origin of many remains was not listed in early cataloguing efforts, collections that were specifically intended to demonstrate the viability of racial classification schemes were more specific.

Quite unlike the vaguer stories behind the origin and acquisition of many of the skeletons in the museum, the largest collection of skulls in the museum, the Hyrtl Collection, consisted of nearly 140 complete skulls bearing relatively detailed—and racially specific—provenance information (the Hyrtl Collection is described in more detail later in this chapter). Following the acquisition of the major Hyrtl Collection, the museum acquired through donation and purchase the "skull of a native of the Sandwich Islands," in addition to a "Negro skull" and a "normal Caucasian skull." Also included in the skeletal acquisitions that followed were a skull listed as "Irish," a "Crow Indian Skull," and the skull of a "Sioux Indian Prisoner." The acquisition of racialized human remains was not limited to skulls and skeletons. Within a series of catalogue entries documenting the acquisition of placenta is the "placenta at full term of a Maori woman (New Zealand)." The entry continues, "Undeniable likeness to monkeys of the placenta surely does not coincide with the intelligence of the race of the Maori."[56] Medical museums were clear in their increasing desire to classify examples of pathologies, but they only occasionally collected *racially*, hoping to help advance the science of racial classification, which was still growing in many anatomical and anthropological circles of the era. While race was not the sole driver in shaping medical museum collections, the practice of racialized collecting had notably permeated medical institutions.

DISPLAYING MEDICAL BODIES

By the close of the nineteenth century, the Mütter Museum's collection grew to the point that its modest exhibition spaces felt tightly cramped. The museum's layout was generally attractive to visitors, but curatorial staff worried about preserving collections in overcrowded exhibit cases.[57] As the collections intended for display grew, not all members of the college were in agreement regarding the utility of the museum. In February 1885, the president of the College of Physicians of Philadelphia, J. M. Da Costa, expressed some frustration about the fact that funds dedicated to the college were intended for the acquisition, study, and display of medical specimens. Instead of creating an ever-larger medical museum, he argued that funds should be diverted toward creating a laboratory for histology and pathological research. Throughout the late nineteenth century and into the twentieth century, some members of the College of Physicians argued that funds would be better spent on research or professional activities of the fellows rather than on the maintenance of a medical museum. The terms of Mütter's bequest, however, were clear.

Only a few years after Da Costa's critique of the value of the museum collection, a new president of the college, D. Hayes Agnew, reported his own impressions of the use of the medical museum. He wrote, "It is gratifying to know that medical students from the different teaching bodies of the city are beginning to avail themselves of the advantages of the very large, varied and instructive collection which our museum contains."[58] The fellows of the College of Physicians of Philadelphia were split as to the utility of possessing and displaying a medical collection. This tension frustrated early efforts to craft complete exhibitions. Nevertheless, the museum continued to acquire and display medical specimens, allowing for the education of physicians and medical students during visiting lectures and special events.

By the arrival of the twentieth century, interest in the Mütter Museum had expanded to the point that it became desirable to host temporary and visiting exhibitions in the adjacent halls. Specimens on display in permanent galleries were relabeled and reorganized in the opening years of the century in an attempt to clearly mark each specimen. With specimens placed under glass cases, they were easily accessible to qualified visitors and could be temporarily removed from their cases to be utilized in a demonstration.[59] Periodically, medical specimens were removed from exhibit cases and utilized to illustrate lectures delivered before the fellows and various medical society meetings held in the same building.[60] Before moving into a new building, the museum comprised just three small gallery spaces.[61] In 1904, the same year the St. Louis exposition featured massive medical and anthropological sections devoted to the human body, the Mütter Museum reported that over one thousand medical students visited the small galleries, touring the display cases with their instructors from nearby medical colleges.[62] Attendance in the expanded museum galleries increased over the next several months. The medical students reported to the curators that the "oral lessons were highly prized." As the students visited the galleries and observed the demonstrations of their professors, the museum reported, their behavior "has been faultless. Perfect order has always been maintained, and not a specimen has been injured or lost."[63] Medical students, professors, and the museum had come to an informal agreement: students could view the rare and useful collections of specimens firsthand, if they behaved in a manner deemed appropriate while they toured the galleries.

At the start of the First World War, exhibition halls at the Mütter Museum began to receive a slightly more diverse array of visitors. Fellows of the college remained the most common, but other respected members of the medical community were also invited to tour the galleries. Though objects remained labeled in complex medical terminology, members of the "laity" were allowed to tour the galleries on occasion. For those who were involved in the operation of the museum, it was a point of pride that the audience for the galleries was increasingly national and international.[64]

Despite the fact the Mütter Museum became increasingly open to the public, the small galleries of the museum hosted a comparatively small number of visitors in relation to larger natural history museums and thus had only a moderate influence on popular American culture. Museum staff complained that the small book intended to register visitors was frequently left unsigned, so the number of total visitors to the galleries was difficult to assess. The museum claimed that the geographic diversity of the visitors increased, yet the number of confirmed visitors for the first full year following the First World War totaled only about 160.[65] By the end of the Second World War, this number increased to over six hundred—a larger, yet still modest, annual sum.[66] These confirmed figures still paled in comparison to the estimated totals, however, and it is likely that a larger number of uncounted visitors toured the galleries.

Displays of specimens consisting of human bodies or body parts in medical museums were in certain ways not wholly unlike displays of human remains in natural history museums, yet medical museums' exhibition halls possessed several notable traits that made them unique. At the Mütter Museum at the turn of the century, specimens could still be easily removed from their display cases in order to be used for close examination in teaching demonstrations. Natural history museum exhibitions, on the other hand, had a greater sense of permanence and were not typically intended for handling. Throughout the opening decades of the twentieth century, the Mütter Museum, unlike the Army Medical Museum, was still primarily geared toward medical students and physicians,

Not only were museums of natural history drawing a much larger audience than did the Mütter Museum; the audiences of major metropolitan museums were more diverse. Despite these differences, medical museums and natural history museums still shared a number of aims, and these shared goals led to both cooperation and competition.

TENSION AND COOPERATION BETWEEN MUSEUMS

For a brief moment in the closing decades of the nineteenth century, several museums and medical colleges attempted to create comprehensive collections of human skeletal remains on a more global scale. Increasingly important in broader American culture, racial theories—including eugenics and even the lingering influence of phrenology—encouraged physicians to submit skulls to medical museums. Although these museums represented different constituencies and possessed slightly different goals, their collections also overlapped, leading to both competition and opportunities for cooperation. While a competitive sentiment between natural history museums had a precedent, the Army Medical Museum and Mütter Museum generally worked together by frequently exchanging duplicate collections. The Mütter Museum, in particular, frequently requested from the AMM examples of pathological specimens showing the effect of bullet wounds on the human body. Meanwhile, the Army Medical Museum submitted to the Smithsonian Institution ethnological material that it had acquired. Despite the general sense of cooperation between these three institutions, examples abound in which museums became territorial. A request made by the Mütter Museum for duplicate crania collections in the Smithsonian was denied; curators at the Smithsonian rationalized their decision by arguing that duplicate specimens of the kind in question should be sent to a region with a dearth of human remains collections—not to a city like Philadelphia, which had become a national leader in both medicine and collections of medical specimens.[67] As part of the effort to build a comprehensive collection, the Smithsonian was unwilling to part with valuable skeletons, at least not without some sort of compensation.

An early interest in racial classification did not entirely prevent the Smithsonian from collecting human remains that showed evidence of pathological conditions or medical practice. On the contrary, collections reflecting these conditions became increasingly important to natural history collections as the twentieth century wore on, but only when these collections were in historical contexts. As anthropologists became increasingly interested in human history, various forms of body modification and surgery became desirable to collect. Between the late nineteenth century and the middle of the twentieth century, however, natural history museums that were engaged in collecting human remains were often occupied by the large-scale project of recording and classifying the "pure" races of humankind, as well as human remains that reflected the mixing of different racial groups. Modified skulls, therefore, were considered of less value than were those that exhibited natural characteristics. For many natural history museums, then, collecting the bodies of individuals whose remains reflected a pathological condition assumed secondary importance until much later. Whereas medical museums sought out these types of collections from hospitals and morgues, natural history museums collected human remains that possessed pathological characteristics only opportunistically from the graves and cemeteries they exhumed.[68]

RACE AND HUMAN HISTORY IN THE MEDICAL MUSEUM

The Mütter Museum, mirroring other museums throughout the nineteenth and twentieth centuries, purchased select specimens while relying heavily on donations to build its collections. The bulk of the collections focused on interest in human health and medicine, but other acquisitions pointed to the prevailing concern with race and human history that existed throughout the period. Earlier museum exhibits had helped spread the notion that races could be understood through racial typologies, and the exhibition of skeletons demonstrated the museum's role in this process to the American public. A skeleton, or parts of the skeleton, could be studied or displayed in a variety of ways, and the rhetoric surrounding its acquisition reflected these varied desires. Further, the collections arriving at the museum often reflected the variety of interests of a physician or the common practice of opportunistic collecting among the educated elites that was occurring throughout Europe and the United States.[69] Three years before purchasing Adam Politzer's collection of auditory apparatus displayed at the international exposition in Philadelphia, the Mütter Museum purchased seventy skulls from Joseph Hyrtl, a physician from Vienna; the skulls were intended to reflect "all the tribes of Eastern Europe."[70] Inscribed on the skulls were the name of the individual, age at death, cause of death, occupation, and religion. Unlike other scholars collecting skulls during the same period, Hyrtl interpreted his collection as evidence of the incredible variation *within* ethnic groups—as opposed to other skull collections, which were interpreted as reflecting the stability of racial groups. Steeped in a devout Roman Catholic worldview of the era, Hyrtl believed that while the external features of humankind responded to environmental conditions, the human mind followed a consistent divine plan. The relationship between the skull and intellectual ability, therefore, was completely random.[71] These conclusions were certainly different from those of people who had studied earlier skull collections, such as Samuel George Morton. Hyrtl wrote of his collection, "Such a collection will never again be brought together. It is easier to get the skulls of Islanders of the Pacific, than those of Moslim, Jews, and all the semisavage tribes of the Balkan & Karpathien valleys. Risking his life, the gravestealer must be largely bribed."[72]

In addition to collecting skulls that reflected racial classification ideas of the era, medical museums actively collected examples of human remains and artifacts that reflected prehistoric or historical examples of surgery or body modification. Race and the study of human history, then, were not simply confined to those scholars working in physical anthropology. In fact, when provided with opportunities to collect specimens thought to illustrate ideas about comparative anatomy, race, or human history, the physicians who contributed to the Mütter Museum eagerly jumped at the chance. Medical museums displayed specimens of animals under the guise of comparative anatomy, but the transitional fossils between various species of human ancestors—gradually appearing in natural history museums—were conspicuously absent.

At the close of the nineteenth century, a small collection of ancient skulls from Peru that demonstrated evidence of surgery confused the distinctions between institutions concerned with medicine and with anthropology. In 1894, the College of Physicians of Philadelphia wrote to the Bureau of American Ethnology regarding a cranial collection that the government wished to discard. The bureau responded to the request by informing the college that the skulls had already been donated to other museums and scholars.[73] The skulls were thought to be representative of the ancient history of the Americas, but they were also of value to those interested in the history of medicine. This was certainly not always the case. Numerous examples of modern medical oddities, such as the *Mütter American Giant*, were of little interest to most scholars of anthropology. Nevertheless, on certain occasions, as with the ancient Peruvian skulls showing evidence of surgery, medical museums and physical anthropologists made competing bids on collections. The Mütter Museum eventually acquired and displayed a series of casts of another set of ancient Peruvian skulls demonstrating evidence of surgery.[74]

Although the exact date of the donation is unclear, S. Weir Mitchell (1829–1914) gifted his personal collection of sixty American Indian skulls to the Mütter Museum.[75] Mitchell was a frequent donor to the museum, having been elected a fellow of the college and serving as the institution's president on two occasions. The collection of Native American crania donated to the Mütter Museum by Mitchell had passed between several physicians, probably in the late nineteenth century, before arriving in Philadelphia.[76] Most of the skulls were collected from graves in Illinois, with a handful of crania coming from other locations in the region, including Missouri and Wisconsin. Many of the crania gathered by the physicians possessed some "deformity," which is the likely reason these particular skulls found themselves in the knapsacks of nearby medical doctors. The supposed deformation of the skulls was created by the natives' practice of cradleboarding their young, resulting in the gradual molding of the skull as the child grows. By the end of the century, scientists who studied the practice had finally realized that such deformities did not pass hereditarily; the scientists had finally pegged the cradleboard as the cause of the oddly shaped skulls in certain tribes. Nevertheless, the exact reason for the flattening of the skull was for a time a matter of debate, and both the medical science and anthropology of the era actively sought to differentiate between the effects of cultural practices and of heredity on the human body. Further expanding on the conclusions drawn from the practice, the curators at the Mütter Museum argued in a memorandum that the development of the cultural practice of cradleboarding could shed light on the prehistory of North America. The memorandum reads, "It is found that the old North American or South American skulls are not deformed; the conclusion is that Indian mothers did not adopt the method of carrying children which brings about the unintentional cranial deformities, until after the American race had lived upon this continent for a long time." The scholars researching the skull collection also concluded after a review of their collections and the available literature, "the American race appeared first in America, at a time when man was developed much higher in Europe than at the oldest time we can trace him there." [77] Once a collection of crania was brought to a medical museum in the late nineteenth and early twentieth centuries, then, it could be interpreted in several ways. The collection had clear implications, the physicians believed, for studies on the subjects of culture, heredity, race, and human prehistory. Race certainly remained the dominant factor in how the museum understood the collections, but narratives beyond simple medical histories were occasionally allowed to creep in. Ideas about race were filtered through letters, correspondence, books, and shared conferences and were encouraged through the acquisition of certain types of skeletons. While medical narratives dominated much of the museum, anthropological narratives of racial classification, cultural habits understood through ethnography, and human history all occasionally crept into medical museum displays.

The medical body in the late nineteenth century and early twentieth century, once removed from the hospital, morgue, or grave, transformed its meaning into one or several of many different intellectual or cultural markers. When brought to a medical museum, a Native American body that showed evidence of deformity became significant to studies of race, heredity, and human prehistory. In many ways, these interests represent the transfusion of ideas flowing between disciplines in an array of museums of the period.

THE MÜTTER MUSEUM AT THE TURN OF THE CENTURY

By the turn of the century, the Mütter Museum held the second-largest collection of any medical museum in the United States, second only to the Army Medical Museum.[78] In 1900, when the American Medical Association (AMA) met in Atlantic City, New Jersey, many participants in the meeting decided to travel the hour-and-a-half train ride to Philadelphia. Philadelphia was the home to fifty hospitals and dispensaries, but the main attraction of the city was the Mütter Museum. Visitors from the AMA toured the modest exhibit halls, where they viewed rare examples of diseased and malformed body parts. An article detailing the visit of the medical students and doctors in the *British Medical Journal* noted that they were joined in their curiosity by members of the public. During one evening, the museum was opened to "entertain" both the visitors from the AMA and the public. The journal noted of the evening, "The interest shown by the laymen in these matters was more than an ample return for the great trouble taken by the College authorities in arranging the evening." It was clear, in other words, that in addition to the learned eye of the physicians, the lower classes, women, and the less educated took something away from the exhibition. "Retrospectively," the physicians concluded, "the lesson of the whole matter is that by such methods the lay people can be brought to a much readier and more wholesome appreciation of the aims of the medical profession."[79]

Reports documenting the nature of the opening ceremonies of the new College of Physicians of Philadelphia illustrate the character of the medical profession at the turn of the century. Speakers predicted that "the new building was destined to witness great things. It would endure to see the end of many diseases of which men now thought with anticipatory dread, and it would endure to witness the passing away of much quackery."[80] Andrew Carnegie, the industrialist and philanthropist, had personally donated $100,000 toward the construction of the new facility.[81] The donation complemented a massive series of donations made by Carnegie that led to the construction of new libraries across the country. Whereas many of Carnegie's libraries were founded for the broader public, the library and museum housed in the new College of Physicians of Philadelphia building was still intended primarily for specialists with concern for public health.

W. W. Keen, a highly regarded professor at Jefferson Medical College in Philadelphia, wrote an extended report on the nature of the new facility. Much of Keen's report, again published in the *British Medical Journal*, focused on the valuable medical library housed in the building. Keen added a few lines on the nature of the museum, as well, however, pointing to the desired use of the displays: "Nowhere can the medical visitor to Philadelphia spend a more peaceable or more instructive hour than in our new building amidst its books and specimens."[82] When the new building opened, four rooms, making up about thirty-five hundred square feet of floor space, were dedicated to museum galleries. The galleries were intended to be inviting and instructive to medical professionals, but Keen added that the building was "open to all the non-medical public who may wish to investigate any subject in which they may have some special interest."[83] The opening of the museum to the public proved to be a turning point for the institution, with eager visitors continually proving the earlier assumptions of the curators to be inaccurate.

Those who led the Mütter Museum were in agreement with Keen's assessment that the move to the new facility would make the collections more accessible. Practically speaking, the new exhibit halls were on the ground floor, meaning visitors no longer had to climb a long series of stairs in order to reach the museum.[84] Ironically, the opening of the new galleries allowed the museum an opportunity to reorganize the museum exhibitions to bring the museum into an anticipated modern era that never actually occurred. Instead, the use of medical museum specimens drastically declined in professional use. Many small medical museums shut their doors, destroying their collections or shipping them to other museums. The Mütter Museum, meanwhile, only became an increasingly popular attraction for visitors—and the public continues to be fascinated, confounded, and educated by these collections.

. . .

The legacy of medical museums is decidedly unique, but their stories are intimately bound up with the expanding projects to study race and prehistory through human remains. Modern technologies have largely rendered the very concept of medical collections functionally obsolete with regard to their use for educating doctors. While preserved medical specimens were exceptionally useful in teaching medical students for a time, they have always suffered from the limitations of preservation; physicians rely on color, form, and occasionally even odor to diagnose a patient—all of which suffer (sometimes dramatically) during preservation for medical museum collections. While these museums were founded with the intention of teaching physicians, they have, in some ways, become public curios of the macabre. Despite the best efforts of nineteenth-century curatorial staff, medical museums never became the center for research they envisioned. One historian has simply argued, "Despite [the] efforts [of the curators] . . . few persons visited the museum; none of the local professors used it in his teaching; and some Fellows dismissed it scornfully as something that survived and grew only because it had an endowed income that could not be diverted to more urgently useful projects."[85] On the other hand, medical museums in the nineteenth and early twentieth centuries did little to attract public visitors beyond allowing the occasional journalist or guidebook author to describe visits to the semiopen galleries.

More recently, another museum specialist described the Mütter Museum as "Philadelphia's strangest museum."[86] The history of medical museums in the United States, however, allows us to make more sense of this supposed oddity and gives us a more complex picture than offered by the disgruntled fellows who dismissed the museum as a failed attempt to construct a medical research center. Subtly, the museum both responded and contributed to an evolving range of ideas in American culture about the human body. Although these collections, as they pertained to racial classification or prehistory, never ascended to the prominence of natural history museums, they reflected the theories of medical collectors to an audience that ranged from physicians to members of the public.

Medical museums originated around the period of the Civil War, intended to serve as engines for the creation of knowledge among medical students. Over ensuing decades, as ideas revolving around public health and hygiene became a matter of public consciousness, medical museums became a more prominent feature in the cultural landscape. At the same time, medical museums adopted a goal to promote the methods of the professional medical community—separating it from quack doctors who promised simple cures in patent medicines. Medical displays, however, were not simply confined to the museum; they were also featured prominently at international expositions. Developments in anthropology—namely, the growing study of race via cranial studies and the study of ancient medicine in the Americas—influenced displays and collecting patterns of medical museums in critical ways.

Although medical displays utilized human remains as tools following the Second World War, new public health concerns dramatically changed the nature of the education.[87] Scattered displays and modestly sized medical museums had failed to fully communicate crucial public health ideas to mass audiences. Nevertheless, the collection of human remains had come to serve a variety of constituencies. Today, the challenges threatening the relevancy of the medical museum seem in certain ways less contentious than do critiques directed toward museum collections that are more firmly ensconced in the study of indigenous peoples. Exhibitions on medical history built on the strength of the collection, and while occasional controversy surrounded the display of medical bodies, the challenges brought forward were less vehement than were later challenges brought against the display of vast numbers of indigenous remains in the bone rooms of natural history museums.

Today, the Mütter Museum maintains a collection totaling about 20,000 objects.[88] Between the mid-1980s and 2007, the Mütter Museum's attendance ballooned from about 5,000 annual visitors to more than 60,000 visitors per year.[89] The museum is now the most visited small museum in the city of Philadelphia, attracting over 130,000 visitors each year.[90] Clearly, visitors are still "burning with curiosity" to view medical oddities, just as they were in 1874, when Chang and Eng arrived at the museum for an autopsy. Over the course of the past few decades, the Mütter Museum has assumed an increasingly significant place in the study of the history of medical science, delicately attempting to reshape aspects of a public health mission while also becoming an active historical research center. Human remains continue to be prominently displayed, and they still draw a respectable audience to the modestly sized institution.

Unlike medical photographs or textbooks, turn-of-the-century medical museums faced the limitations of both time and space. Medical photographs or textbooks could be circulated to physicians away from major urban centers, whereas the physical space of the museum was limited to those physicians who were either visiting or training in the city where the museum was located.[91] Despite this, however, the medical museum did not become obsolete and forgotten overnight. Medical students and the public maintained a level of curiosity regarding the specimens on display at the Mütter Museum, and they continued to visit exhibition halls.

Even as medical museums increasingly became vehicles for popular education and public health in the early twentieth century, the spotlight on displaying human remains in the United States shifted away from medical displays and toward an exhibit hall being built for the 1915 Panama-California Exposition in San Diego, California. Although the modest medical displays of bodies featured at the Mütter Museum in Philadelphia would have been remarkable to the average visitor and pragmatic to the medical professional, the displays planned for San Diego were to be quite unlike anything audiences in the United States had ever seen.

CHAPTER 4

The STORY of MAN THROUGH the AGES

In 1915, with war bleeding across Europe and making its way around the globe, two competing world's fairs opened in California. San Francisco hosted the Panama-Pacific Exposition, and San Diego built the Panama-California Exposition. The San Francisco fair, both larger in size and officially sponsored by the federal government, attracted nearly nineteen million people. San Diego lost the competition to host the official fair but nevertheless created an event that attracted over three and a half million people. The fairs were largely celebratory and bright, with buildings bathed in warm colors and nostalgic portraits of California history, showcases for the latest proclaimed achievements in science, and an occasion to buy and exhibit fine art.[1]

A major attraction in San Diego, and arguably the exhibit at the fair with the most lingering influence on American ideas and culture, was a new exhibit called *The Science of Man*. The exhibition ultimately begat a new museum entirely—the San Diego Museum of Man—but arguably its greatest significance was the manner in which it broke open race and prehistory as themes for audiences in the United States. The exhibit used both art and bones to display racial classification theories. Almost bursting into the exhibition were dramatic and emerging examples of the human past. These themes were presented through a new intertwining narrative of artistry, human drama, and actual human skeletons. If the narrative drama surrounding human evolution had the makings of a compelling story, the scientism dominating the remainder of the exhibition was presented in almost total isolation, lacking any real context or true narrative. Exhibits on race and human evolution at the Panama-California Exposition were, at the time, the most complete exhibition on the natural history of humankind ever shown in public.[2] The founders and early observers of the displays and resulting museum credited the physical anthropology collections, consisting of a unique collection of skeletons and mummified remains, as an important lasting legacy of the fair.

Also significant was the introduction of artistic narrative into the representation of human evolution—a technique that was becoming increasingly central for museums in the United States, especially since they lacked original prehistoric or paleoanthropological specimens from Europe, Asia, or Africa. New discoveries and increased scientific emphasis on prehistory and evolution pushed curators to enrich and dramatize the storytelling behind expanding exhibits. Natural history museums were already in the midst of a major transition, moving from Victorian-era displays of endless rows of glass cases to more creative methods that embraced contextualizing narratives. With regard to displays of human remains in museums in the United States, the San Diego exhibition hall *The Story of Man through the Ages* marked shifting exhibit methods—an imperfect innovation that has continued to influence museum exhibitions on prehistory and human evolution to this day.

Anthropological presentations at fairs and expositions in the United States up to the time of the California fairs typically focused on material culture, repeatedly serving as an opportunity to bring together massive numbers of archaeological and ethnographic objects for temporary display. Often, these newly acquired collections were packed up after the fair, and museums around the world vied to purchase the best collections when fairs concluded. Occasionally, as in Chicago years before the San Diego exhibits emerged in Balboa Park, expositions in the United States were utilized as occasions to create new museums. Previous fairs in Europe, too, such as in Paris (1878) and Dresden (1911), had included popular exhibitions on "Man," while the World's Columbian Exposition (Chicago, 1893) highlighted anthropometry as a growing field.[3] The fair in San Diego was the first to truly place human remains and the new science drawn from human remains at center stage. Indeed, this fair's anthropology exhibit became an attraction at the fair—apart from its elegant Spanish Colonial–style architecture and lush fairgrounds. Racial science explained the human differences on display in San Diego, and in this context bodies conveyed to visitors how scientists arrived at racial characterizations.[4] The exhibit embedded ideas about human evolution and prehistory in the displays, topics that had received comparatively little attention in previous exhibitions. While the exhibition presented some individual variations within populations or "races," the main theme of the exhibition was demonstrating racial typologies based on research taking place in museum collections of human remains. The exact details within racial theory that were accepted by museums in the United States continued to shift, but positivist scientists largely accepted human races as both distinct and determinable. Visitors walking through the exhibit hall learned how studies of human remains, combined with the anthropometry of the living, shaped ideas about humankind. The exhibit postulated boldly that new discoveries were upending existing ideas about evolution. The displays also showed dramatic examples of pre-Columbian skeletal remains, captivating those who viewed them. Ultimately, however, while the exhibit was determined by many observers to be a resounding success in its shaping of popular thought through the lens of physical anthropology, it clung to the supposedly scientifically rigid racial classification theories emerging from bone rooms across the United States. While the displays were influential in the crafting of other displays in later decades, the ideas presented about race science seemed quickly outdated.

The Story of Man through the Ages exhibit hall featured dramatic displays of human remains, complemented by artistic works portraying the racial diversity of humankind defined in terms of "racial classification." The Panama-California exhibit differed from previous displays in two significant ways. First, the displays introduced many new ideas of human evolution at the start of the exhibition, a topic largely ignored or unknown by creators of earlier displays. This venture also differed from earlier exhibits on humankind in its transparent reliance on the study of the human body. The remains of dozens of individuals were exhibited: bones with simple labels visible under simple glass cases. *The Story of Man through the Ages* brought together ideas and examples drawn from thousands of specimens collected from around the globe, and it worked to blend three major topics: evolution, prehistory, and race. Despite the seemingly modernized collection practices informing the presentation, the displays in San Diego actually stood in contrast to a growing relativism—or notions about the racial and cultural equality of humankind—in the anthropological community, which was moving toward the idea that differing cultures are equal in complexity/sophistication, rather than seeing them as moving along a ladder from primitive to civilized.[5] Influenced as much by Lamarck as Darwin, art dramatized the evolution of humanity, emphasizing the growing size of the brain cavity and humankind's increasing talent for making tools. As ideas about the equality of cultures entered the discipline, positivist ideas regarding racial classification were pushed to the verge of collapse. Anthropological exhibitions, such as the one in San Diego, that centered on human remains continued to separate and classify the races of humankind. The scholars who mounted the exhibition remained convinced that new discoveries related to the study of human evolution contributed to understanding the modern human species, particularly the emergence of racial difference.

CRAFTING EXHIBITIONS FOR SAN DIEGO

Edgar Hewett, an archaeologist noted for work in the American Southwest, was hired by fair organizers to craft exhibitions on archaeology and ethnology. Hewett earned a reputation as an effective and clear teacher, and he had also become known for successfully helping to lobby for the American Antiquities Act a decade earlier. His skills as an effective communicator and academic politician made him ideally suited to lead the effort to create displays for the fair, which might be nationally and internationally scrutinized.[6] While many anthropologists admired Hewett's ability to advocate for archaeology politically, some resented his growing stature in the field and envied his leadership position in San Diego. Hewett, for his own part, distrusted his new colleagues in San Diego and turned to his counterparts at the Smithsonian for help organizing the exhibition. Hewett hoped the fair could serve as an opportunity to build a new museum from the ground up—a museum that he might someday lead.

Soon after being hired to organize an exhibit, Hewett made contact with William Henry Holmes and Aleš Hrdlička at the Smithsonian.[7] Enjoying a generous budget stemming mainly from wealthy San Diego boosters, Hewett worked with the Smithsonian curators to complete a plan for the exhibitions early in 1912. In March, the Smithsonian officially agreed to a plan allotting $27,000 toward Hewett's budget for an exhibition in physical anthropology.[8] Another $5,000 was allotted to Holmes, who agreed to use the funds to craft displays on the "mining and quarrying industries and the stone working of the American tribes."[9] The Smithsonian had routinely mounted exhibitions of American archaeology and ethnology on this scale, but the amounts dedicated for displays in physical anthropology were unprecedented.

The National Museum's rationale for such an agreement was clear. Previous fairs reported increasing public interest in the comparative study of the human body, and the Smithsonian had the expertise to guide such displays. The fair also presented an opportunity to gather more remains for its growing bone collection, as the budget included funds for travel to collect bones. The secretary of the Smithsonian Institution, Charles D. Walcott, as if writing Hrdlička a blank check, wrote simply, "Your work in this connection will afford an important opportunity to promote the interests of the National Museum."[10] Holmes, the quietly clever, well-connected, and dignified scholar who now served as head curator of the department of anthropology, wholeheartedly supported this idea. Holmes wrote the Smithsonian administration, "It is my feeling, that if carried out, the work planned will not only prove of great importance to the Exposition but that it will be of very especial benefit to the National Museum." Holmes continued, "I am anxious to see Doctor Hrdlička and his Division stand at the head of this branch in America and for that matter in the world, for the study of the race and race interests, must, I believe, grow to much greater importance in the near future."[11] Although Hrdlička eventually collected valuable information on human evolution and prehistory in the course of building the exhibit, Smithsonian participation in the exhibition was initially couched in terms of racial science. Despite his struggles to create any large-scale or permanent exhibit in Washington, DC, circumstances created a perfect storm that allowed Hrdlička to exhibit ideas as made manifest through both bones and art in San Diego.

In a memorandum sent to the museum's assistant secretary, Hrdlička pointed to building collections as a desirable outcome in the plan to create new exhibits in San Diego: "Additions of highly desirable skeletal material from regions and races which are but poorly or not at all represented in our collections [*sic*]. This material will greatly enhance the study and also the exhibitions value of our collections, and there are no means in sight of acquiring it otherwise."[12]

Just days after the Smithsonian approved participating in the San Diego fair, planning for an expedition to St. Lawrence Island, Alaska, began. Two scholars, working for both the museum and the fair, were to take photographs and collect valuable linguistic data. One scholar was also assigned the additional task of bringing back a "collection of skeletal material."[13] With that assignment, the process of building a new museum of humankind for the city of San Diego, based centrally on the research and display of human remains, began. Unlike in the Smithsonian and within anthropology more generally, where physical anthropologists of the era believed their science was being marginalized, the study of human remains was to be central in the new museum in San Diego.

The early plans for the displays were both broad and vague. Plans included sections on "The Evolution of Culture" and "The Native Races of America," in addition to "The Physical Evolution of Man."[14] Hrdlička was successful in gathering a large and growing collection of bones for the Smithsonian, but he failed to convince museum leaders to set aside the funding required to create new physical anthropology displays. Hrdlička also failed to convince organizers of the St. Louis Exposition in 1904 to create large-scale displays of any kind on the subject of physical anthropology.[15] The chance to create displays at the San Diego fair greatly appealed to Hrdlička and his sizable ego. He took on the assignment to collect material for the fair with fervor, and his obsessive ambition pushed him to plan for displays of a massive, global scale.

COLLECTING BODIES TO DISPLAY

Early in the planning stages of the fair, it was agreed that the bulk of the skeletal material Hrdlička collected would become the property of the Smithsonian Institution, with a smaller allotment to remain in San Diego for the new museum, following the fair.[16] This framework for distributing collections followed standards that the Smithsonian hoped to adopt for future collections: the National Museum obtaining the best or most desirable specimens, with representative (or "duplicate") selections of materials sent to other locations—in this instance, to the San Diego fair and later to the museum in the same city. Hrdlička first hired a pair of modelers to create busts to illustrate racial types in different parts the globe. Artists were chosen to closely follow the direction of scientists in the creation of models of men and women of particular races or of distant human ancestors. Scientists instructed artists to depict individuals on the basis of exact measurements of living humans or skeletal remains. Hrdlička then hired anthropologists to collect skeletal material from central and southern Europe. Another anatomist, Philip Newton from Georgetown University, traveled to the Philippines, charged with the task of collecting Negrito remains. Newton's expedition, supervised and supported in part by the U.S. Army, resulted in the desecration of a large number of graves, dug open to collect materials for display in San Diego.[17] Judged simply in terms of collecting skeletons, the expedition was considered a resounding success. Acquiring human remains did not come without challenges, however, as war and resistance from people on the ground proved difficult to overcome.

Hrdlička himself undertook an incredible series of expeditions, collecting in Europe before traveling to Siberia, Mongolia, and then Peru in an era when such travel was still strenuous and difficult.[18] Boxes of materials, most filled to the brim with human remains, poured into the Smithsonian during Hrdlička's extended travels.

Before his departure abroad, however, Hrdlička's planned Siberian expedition to collect skeletal materials and measurements encountered roadblocks. Upon learning of his planned expedition, the Foreign Office of St. Petersburg informed the Smithsonian that the plan could not be approved until more detailed itineraries were submitted and approved.[19] Eventually, through working with U.S. government officials, Hrdlička gained access to Siberia. This experience was not an isolated one; in the early years of the twentieth century, governments around the globe became increasingly protective of antiquities found on their soil. Growing nationalistic concerns fueled legislation to protect antiquities, keeping the best antiquities, relics, and remains for national museums and only allowing certain material to be sent abroad. Hrdlička, however, distinguished between the value of antiquities and the scientific value of human remains. He wrote about his plans to explore Peru, "The object of my visit will be restricted to observations on the living and collections of skeletal material. There will be no excavations for or collection of antiquities."[20] Whereas antiquities could be sold on the art market, he argued, human remains held little value outside of studies on race, prehistory, or medicine. In this context, Hrdlička viewed, the value of the body was almost solely in its use as a scientific or teaching instrument (this said, he was not above inquiring about the price of certain remains for possible purchase). The term *antiquities* had at various times been utilized to describe both artifacts and human remains; Hrdlička distinguished skeletons and mummies as separate specimens valued only as objects for science.

Others also faced similar obstacles to collecting skeletons. Hiram Bingham, the Yale University archaeologist credited with discovering Machu Picchu in 1911, two years later wrote to William Henry Holmes about the changing conditions for removing archaeological materials from South America. Bingham, who was thought to have tried to establish an academic monopoly for Yale in Peru, told Holmes that he believed Hrdlička was going to face new challenges in South America during his planned expedition: "He is going to have, I am afraid, considerable difficulty in getting permission to investigate graves and export bones." He continued, "Although the material which he is after is of no particular value to the Peruvians, and although they would not know what to do with it if they had it, the very fact that he is willing to come such a long distance, and spend money in securing it, is sufficient proof to them that the material that he is after is material that they ought to keep in the Country."[21] Although Bingham and other archaeologists of the era were often willing to work within the confines of new international regulations, they typically despised the notion of asking permission to work on materials that host nations seemed uninterested in either studying scientifically or preserving. Though archived records suggest that archaeologists of this period seldom considered indigenous rights to control the remains of their ancestors, archaeologists did gradually come to respect national regulations regarding the ownership of antiquities. Skeletons stood apart, scientists continued to plea, arguing that a lack of value on the black market meant that their only value as objects was as scientific specimens.

Scholars interested in human remains and ancient antiquities had reason for concern, though the problems were often far different from what they anticipated. What Hrdlička found when he arrived in Peru was shocking. Ancient cemeteries lay ransacked and looted. Artifacts were removed and littered across the surface of the ground. Looters ransacked thousands of graves from ancient cemeteries in the search for gold and relics. Ancient human bones were simply tossed aside, left to bleach in the hot sun. With the graves unturned and bones now laid bare on the surface, Hrdlička was able to collect an unprecedented number of pre-Columbian crania. The circumstances also afforded him the opportunity to quickly find and collect rare examples of skeletons that showed evidence of disease, trauma, and surgery, which were uncovered and left behind by looters who were primarily interested in artifacts with potential to be sold on the illicit art market.[22]

Hrdlička's interest in Alaska, Siberia, and China was due in large part to his interest in the history of the peopling of the Americas. He observed that among the people of the parts of Asia that he planned to explore "are found physical types in every respect identical with the American Indian." Simply stated, American Indians appeared to share certain physical features with people in Asia. He continued, "The object of my trip is to trace, in a preliminary way, the remnants of the stock of people from which in all probability the American race branched off, a problem which is becoming one of the most important subjects of research in American anthropology."[23] Hrdlička's interest in Peru had a slightly different rationale. In writing to the State Department, Smithsonian secretary Charles D. Walcott explained his intent: "The main purposes of his studies and collection will be to ascertain the distribution of various physical types of man in Peru, and the study of the diseases to which these native populations were subject before their contact with the Spaniards."[24] Despite Hrdlička's fears to the contrary, the Peruvian government welcomed his collecting of human remains for the exhibition.[25] At the time, the claim that scientific expeditions were interested only in human remains worked to assuage certain governments, especially colonial governments that were largely uninterested in indigenous attitudes toward death and burial. Despite being granted formal access to collect skeletons in Peru, Hrdlička told the *New York Times*, "The opportunities for getting prehistoric skeletons in the rich burial grounds and ruins of that country will have practically vanished four or five years hence." The difficulties that scientists in the United States faced in collecting these remains would no doubt allow looters to ravage the ancient burials, the paper reported.[26]

Despite setbacks, the Smithsonian considered Hrdlička's global expeditions a success. In Europe, he examined the most significant ancient human remains in museums, a claim that no other American could make at the time.[27] He used the opportunity to have new casts, or detailed replicas, made. Hrdlička noted, "These casts will supplement our collections which are already richer in this line than any other on this continent." Although foreign museums could request that original skeletons be cast at any time, no guarantees attested to the ability of the museum to create accurate, detailed copies. Hrdlička believed that by making casts of skeletons in European museums himself, the Smithsonian could ensure the accuracy of the casts in various comparative studies. Casts were considered easily measured and could be used repeatedly for study. A student might view the cast of a rare fossil or skeleton whose original might be housed in a museum thousands of miles away, comparing it against material found in the bone rooms of the Smithsonian. In addition to casts, Hrdlička's travels in Europe yielded a collection of artistic representations of ancient human forms in plaster, as well as a modest new collection of ape skulls and skeletons. These, too, might be judged against the many thousands of human skeletons already in the collection. Before Hrdlička concluded his work in Europe, the curator organized excavations in Bohemia and the Ukraine, resulting in collections of early historic and prehistoric material that was then extremely rare in the United States. Copies of the casts and renderings were sent to Washington and San Diego, building the Smithsonian collections while also preparing for the San Diego exhibition. Meanwhile, remains from Newton's expedition to the Philippines added to the continued stream of bones arriving at the Smithsonian.

Later in the summer of 1912, Hrdlička left for Russia and Mongolia, where he both collected and scouted regions for previously unknown collections of remains. In Mongolia, for example, Hrdlička leveraged the assistance of "the Russians and some

Cossacks" to gather 215 skulls of Mongolians and 15 skulls and a single complete skeleton of a Buraits individual.[28] The majority of these collections were eventually deposited in the U.S. National Museum; however, curators first selected certain materials for display at the Panama-California Exposition in San Diego.

Casts and artistic renderings of ancient humans were valued, but Hrdlička argued that the bones he collected in Asia held data to test another idea. The most significant science stemming from observation of modern human remains from East Asia, it was argued, was the comparison of Asian physical types to modern-day American Indians.[29] The Smithsonian wanted to test the idea that ancient humans had crossed the Bering Strait land bridge to the Americas. To do so, the museum wanted to compare Asian populations to the indigenous peoples of the Americas. The museum therefore desired thousands of skeletons from all across Asia. These skeletons from Asia could then be compared to the already massive collections that the museum possessed from the Americas. Evidence collected by Hrdlička during his efforts to bring together material for the Panama-California Exposition informed his subsequent efforts to describe the relations between the populations of East Asia and the Americas. This included Hrdlička's controversial claim of a relatively recent arrival of American Indians in North America. In this case, as in numerous others, the impetus to collect provided by a new exhibition resulted in permanent collections that were important for future research.

When Hrdlička returned to the Smithsonian, he busied himself with organizing expeditions that were supervised mainly by individuals he had met recently while in Europe. Expeditions to British East Africa, Australia, and South Africa were hastily planned; however, all three were soon canceled. The slated leader of the initial expedition to British East Africa fell ill, and paperwork for the Australian expedition failed to be processed in time to make the expedition possible.[30] Plans were again changed when a young scholar whom Hrdlička met in Prague, named Adalbert Schück, agreed to collect in South Africa and British East Africa. The plan allowed for Schück to collect skeletal materials and take photographs of indigenous communities in Africa, with the hope that new pygmy material would directly add to the science of classifying contemporary African Americans in the United States.[31] Upon hearing of the outbreak of war, Schück reported to government officials in British East Africa, presenting his papers and letter of introduction from the Smithsonian. Schück, a native of what was then Austria, was treated as a suspected spy and arrested.[32] Eventually, he was allowed to leave the country, but he was not permitted to ship what he had already found to Washington.[33] Although the hardships encountered by Schück in collecting human remains abroad were not entirely representative, they do point to various challenges experienced by anthropologists who hoped to collected human remains for research and display in the United States, especially those working abroad at the outset of the war.

CONSTRUCTING RACE AND HISTORY

Hrdlička described *The Story of Man through the Ages* as an effort to "bring together a comprehensive, instructive and harmonious exhibit relating to the natural history of man."[34] Despite the title, a narrative describing human evolution was only embraced to a rudimentary extent—apparent to visitors mainly through artistic reconstructions of humankind's earlier ancestors. The exhibition was essentially a demonstration of the many scientific facts that the Smithsonian believed it could draw from its growing human remains collections, as well as the collections of the fossils and bones of human ancestors housed in museums abroad, although the vast majority of visitors would not have understood it as such. Significantly, a reliance on artistic representation of human remains worked to introduce new forms of narrative into exhibits on the subject of human evolution—a major turning point in how museums in the United States represented human evolution. Subtle introductions of bronze busts and the story of humanity's evolution notwithstanding, the major feature of the exhibit was the large number of human skeletons and mummies—representing just a portion of the entire number of remains added to permanent bone rooms in San Diego and Washington. Taken together, artistic recreations, casts, and actual human remains in the exhibit intended to promote and advance the science of physical anthropology as a discipline. Some journalists dramatized, almost fictionalized, their own accounts of human history on the basis of the bones displayed in San Diego, but the narrative surrounding humanity's evolution and development into supposedly distinct races was far from explicitly presented. The organization of the exhibit reflected the changing features of physical anthropology at the Smithsonian, and elsewhere, around the time of the First World War. The displays that ultimately debuted in San Diego were in a centrally located part of the fairgrounds. The opening room featured natural light streaming in from windows built into high ceilings above, onto both bronzes and bones.

The opening room, generally following the original plan for the building, featured a working anthropological laboratory, library, and desks that potentially be converted into a podium for lectures.[35] Lecturers spoke on subjects ranging from anthropometry to evolution, mirroring the subjects of the four following rooms. Visitors to the fair could observe an anthropologist taking cranial measurements of brave volunteers or explaining the use of particular scientific instruments. A lecture might conclude by informing visitors of the significance of the whole exhibition. Hrdlička's implicit arguments espoused individual variation within constructed racial groupings, though at the same time, the exhibition posited the evolutionary development of particular racial groupings. Visitors learned that racial types had evolved over time from distant common ancestors. Individual variation, as interesting as it may have been, was subverted by overarching theories of racial classification. Hrdlička developed his exhibitions to express the notion that while evolutionary change occurred, it did so within racial groups, not across them. Lectures and demonstrations in the opening room introduced many of these ideas to visitors before they entered the main galleries. Though the exhibition carefully avoided positioning certain races as more evolutionarily advanced than others, it seemingly argued that certain racial groups maintained some primitive features. Instead of presenting a clear narrative arc on racial history, the exhibit focused instead of identifying races; the human body, it was shown, just as with animals, could be neatly identified, organized, and classified as a specimen. Instead of detailing the relationship between the peoples of Asia and the Americas through the story of the journey of modern humans across the Bering Strait, the exhibit hinted at their close relationship by comparing the features of their bodies.

As visitors moved into the second room, they found displays on human evolution. Natural light streamed down to the displays from the ceiling above. In the center of the room was a series of ten dramatic busts. The busts were something of an experiment. Although artists and scientists had worked together for centuries to create detailed illustrations of natural history specimens, accurate illustration of scientific ideas related to physical anthropology seemed to lag behind. When displaying ideas drawn from bone rooms, early exhibitors instead relied heavily on the actual remains themselves to convey the ideas behind the exhibition; casts and detailed scientific illustrations only occasionally stood in for authentic specimens. Artists and journalists with little scientific training or actual knowledge of recent discoveries popularized the simplistic notion of the "missing link," in both the United States and Europe. With new bones and fossils gradually emerging from Europe, Asia, and Africa, it was becoming easier to get lost in the complex web of human prehistory. In order for these discoveries to be communicated effectively, visitors needed a story to latch onto—something to identify with on a human level. The series of busts in the second room, in a manner quite unlike previous exhibitions of ideas drawn from human remains, worked to draw out a particular narrative of human evolution as told through bones. The earliest figures possessed exaggerated, simian-like features—including one bust of a mother dramatically protecting her small child. Many of the statues, including some of the busts intended to represent our earliest known ancestors, included replica tools. One bust casually looks toward the visitor, even as the statue features a recently killed pig resting over the man's shoulder. The later busts looked increasingly human-like, with ascribed features appearing both dignified and strong. Humankind was portrayed as undeniably advancing through the ages, standing upright, wearing decorative jewelry and clothing, and using tools of stone and wood.

Sitting below rows of simple wooden rafters in the temporary fair building, the exhibit introduced visitors to a basic narrative of human evolution. The story of humanity's rise and prehistory was told not just in casts but also through vivid sculpture. What seemed new in this exhibit was the translation of the ideas from bones into art. The artistic skill demonstrated by the sculptures was not exceptional, yet some of the sculptures were effective in lending a living quality to some of our most recent human ancestors. Although charts and maps hung on the walls, gently tilting forward toward the visitor, the information behind the sculptures was limited. While the charts and maps were intended to contextualize the discovery of original remains, the exhibit offered little about the lived existence of each of the species. Any narrative of human evolution offered in the room was stunted by its incomplete nature, voids that future exhibit designers would seek to fill in teaching even larger audiences about human history.

In addition to the busts that provided some semblance of human narrative behind human evolution, visitors to this exhibition encountered other new types of displays. For the first time, museum visitors in the United States viewed exact replicas, or casts, of the skeletons of prehistoric humans. Many of these replicas had been recently cast for the fair by European museums holding the originals. Combining the casts with the busts was clearly intended to give the visitor a greater sense of the science of physical anthropology through the use of artistic representations of human ancestors. The exhibition's catalogue emphasized that the scientific enterprise had advanced over the course of the past thirty years: "Skull after skull as well as other bones of the skeleton have been discovered, and under conditions which enable men of science to establish their great age beyond a reasonable doubt."[36] These artistically created busts, then, were framed by the science of collecting and studying bones.

Notably, the exhibition instructed visitors that the primitive features one might observe in the reconstruction of our human ancestors might also be found in certain living human populations. The exhibition did not imply which human populations were primitive and which were more advanced, yet one might imagine the direction such statements led the majority of visitors of the era. Visitors could view large maps showing where the original specimens were discovered, and charts were displayed showing how archaeologists estimated the specimens' ages on the basis of the context in which they were found. Illustrations of ancient humans lined the walls, and benches were spread throughout for visitors to rest and ponder the long course of human evolution. In the center of the second room was a dramatic set of ten busts, displaying in vivid detail, at eye level, the artistic reconstructions of various species of human ancestors. A Belgian team, an artist named Louis Mascré (1871–1927), who was supervised by the museum scientist Aimé Rutot (1847–1933), crafted busts for the exhibition. It is unclear the extent to which Hrdlička, Mascré, and Rutot influenced each sculpture; however, the reminder that scientists and artists were collaborating was repeated throughout writings on the busts. Also central to the creation of the sculptures was the modeling of human ancestors from Europe and Asia on the basis of studies drawn from human remains housed in museums. One description of the busts pointed to the connection between the artistic renderings through sculpture and actual human remains: "These models are constructed from the actual skeletal remains, and the decorations and implements are exact reproductions of those found with the bones."[37] In some sense, the exhibit made the connection between these statues and original remains clearer than it did in making connections to narratives of evolution. Sculptures revealed hominid figures making tools, carrying animal prey, and gazing off into the distance as if in the early stages of modern human thought.[38] Hrdlička promoted the busts as "striking and interesting."[39] Concluding the displays in the second room was a series of crania of contemporary primates, moving from the lemur to an example of a modern human. Visitors walked through the center of the room to view the sculptures before examining replica skeletons and skulls that filled the cases at the other end of the room.

The third room introduced visitors to a scientific perspective on a familiar idea: aging. Human aging as a subject, even more so than the subject of evolution, was heavily racialized in *The Story of Man through the Ages*. The room was, in fact, an exploration of comparative racial theories of North America. Statues of human busts—exact copies of human bodies from the chest to the top of the head—lined three separate groups of cases. Frozen in time, these busts were presented less as a life narrative than as representative specimens of life stages within each particular racial category. The busts were separated by both race and gender and then arranged from youngest to oldest. Each of the three sets contained a total of thirty busts—fifteen males and fifteen females. The "Old

Americans" or "thoroughbred" whites who had occupied the continent for three or more generations were compared to American Indians, who were represented by a line of small statues and busts depicting Sioux individuals from birth to old age.[40] In one set of cases, the individual Dakota were collectively intended to represent the vast and diverse range of American Indians across North America, with hair and facial features changing subtly from one end of the case to the other. Each figure possessed a serious and straightforward expression, as though permanently focused on the visitors walking by. A third series, representing "the full-blood American negro," completed the room and included a bust of a woman thought to be 114 years old.[41] The Smithsonian boasted of the display, "These series, which required two and one-half years of strenuous preparation, form a unique exhibit, for nothing of similar nature has ever been attempted in this or any other country."[42] Observant visitors would likely have noticed the differences between individuals of the same age, but their grouping by race underscored racial classification as the key idea of the room. These busts represented more than just individuals; they represented the American Indian as a whole, progressing, like all races, inevitably into old age, decline, and death. Science, so it seemed, was not without a morbid dose of reality.

Exhibit designers hoped visitors would pick up on the theme of racial difference through close observation of the busts across different life spans. Though the aging process seemed to be similar across racial spectrums, the exhibition emphasized that "while remarkably alike in all parts of the inhabited globe, [the busts] show nevertheless racial and environmental variations." The portion of the exhibition that was most firmly embedded in ideas of racial classification emphasized something of human unity while simultaneously privileging the notion that races varied widely based on ancestry and environment.[43] Writers who toured the exhibition read the displays as either showing similarities in the aging process or providing reason to believe that humans vary wildly across a racial spectrum.[44] One magazine article read, "The variations between the so-called white, black and yellow races, is very marked both in facial characteristics and bone structure; and the vast differences between Indian, Eskimo, Mongolian, Negro and other peoples are shown by means of casts taken from life."[45] Evidence for essential racial difference was understood to be in our very bones—evidence that could be used to organize race, throughout each stage of life and influenced subtly by environment and individual ancestry.

Entering the fourth room, the exhibit's emphasis on race continued as visitors viewed directly the whole of the science of racial classification. Walls were filled with two hundred photographs intended to provide "racial portraits" of people from around the globe. Over one hundred facial casts filled glass cases, including a collection of facial masks of Bushmen, considered "especially rare and valuable."[46] Yet more artistically crafted busts surrounded the room, with male and female individuals representing the whole of several groups from around the world.[47] The busts in the room were created by an Austrian artist named Frank Micka.[48] Hrdlička described Micka, who had earlier immigrated to the United States, as "one of the best modelers in this country." Micka was an above-average artist, but his background as a Czech national influenced Hrdlička's high opinion of him as an artist and illustrator of scientific facts. The busts included some detail, including hints of skin tones and the traditional clothing of the groups that the sculptures were intended to represent. Nevertheless, each pair of busts was more or less presented in distinct isolation—standing alone for massive populations of humanity. Hrdlička continued by describing the casts that Micka produced as being "actual casts of the face and body of the several subjects with the help of careful measurements." This resulted in what Hrdlička believed were "accurate racial records, the value of which will increase with time."[49]

Hrdlička's use of the phrase "racial records" is revealing as it provides insight into his belief about the development of collections—including mummified and skeletal remains, as well as casts and sculptures based on detailed measurements—that resulted in a virtual snapshot of global racial anatomy at a time when he feared that certain racial groups were quickly vanishing. Racial mixing, genocide, disease, and environment all might play subtle but critical roles in shaping the human body, yet these barely registered as themes in this exhibition.

The final room was the most grisly. At first glance, the room almost appeared to be a systematically organized mass grave. Unlike burial grounds, however, these bones—mostly skulls—were laid out in patterns in flat-top glass cases. A series of charts and maps circled the room, illustrating in vivid detail the most common causes of death for humans in different areas of the globe, but the primary focus of the room was the skeletons recently brought to the United States from Peru by Hrdlička himself. Examples of disease were "illustrated extensively" with actual human remains from the Americas. The center of the room contained a series of flat-top cases containing "many hundreds of original specimens, derived principally from the pre-Columbian cemeteries of Peru, show[ing] an extensive range of injuries and diseases, such as have left their marks on the bones."[50] Indeed, the vast majority of the remains on display featured obvious and dramatic fractures or cuts or the slow decay of some awful disease. The curators who organized the displays believed that visitors would find this series of cases to be of great interest, due in part to the fact that some of the individuals had actually recovered from horrific injuries or from a disease that had afflicted them during life. In a report describing the final layout of the exhibition, the Smithsonian specifically noted the sixty skulls illustrating pre-Columbian surgical techniques.[51] Skulls that showed evidence of trepanation might have shocked visitors while at the same time attracting a morbid curiosity tied to an interest in the prehistoric and the exotic.[52] A further description of the exhibition dryly stated, "The people fought with clubs, maces, and slings, and the resulting wounds of the head, if not fatal, left generally impressions of bone, which must have given rise to serious symptoms."[53] Even to the untrained eye, the smashed, deformed, or partially healed bones quite likely proved captivating, especially in a time when professional mortuary services and health care were increasingly separating the average person from death and dying. Recognizing the compelling nature of the displays, Hrdlička wrote, "In many instances the injuries are very interesting, both from their extent and the extraordinary powers of recuperation shown in the healing; while among the diseases shown on the bones there are some that find no, or but little, parallel among the white man or even the Indian of to-day."[54]

Skeletons in the newly acquired collections that were split between San Diego and Washington, DC, included examples of syphilis, osteoarthritis, fractures, dislocations, and natural mummification.[55] Syphilis, thought to be particularly interesting to visitors because it was a familiar disease, had a gruesome effect on the body, leaving visible scarring, deformation, and deterioration on the bones of those who suffered from the disease.

While ideas about both race and prehistory were apparent in the displays, the notion that humanity evolved into distinct races was a recurrent theme throughout the entire exhibit. Aging, appearance, and ability were directly tied to race, and the natural history of humankind, or the science of physical anthropology, held the key to understanding the varieties of human difference. While racial difference was a familiar idea to the public in the United States by this era, the opportunity to view rare casts and actual human remains brought to it heretofore unseen level of attention. Individual variation, aging, and the effect of cultural manipulation may have an effect on the body, but race remained the all-important category through which physical anthropologists defined humanity. Although the exhibition had opened with the beginnings of a story, the narrative faded as the exhibition moved into the comparative study of race for the two largest rooms of the exhibition. Despite flawed presentation and even in the dizzying context of the fair, the exhibit attracted a large audience—cementing plans to create a new Museum of Man.

REACTION TO DISPLAYS

As visitors strolled through displays in San Diego, Aleš Hrdlička was elated to receive letters from former president Theodore Roosevelt, who was curious about the archaeological finds shown to him when he traveled throughout South America. Roosevelt wrote out of curiosity, asking for more information regarding the peopling of the American continents.[56] Hrdlička was impressed by Roosevelt's interest, and his response sheds light on his thoughts about the visitors to his galleries across the country in San Diego. Hrdlička wrote Roosevelt, "Such healthy, critical interest you are taking in the subject of man's antiquity on this continent is a genuine encouragement. What we usually meet with, and that even on the part of intelligent people, is either a blind acceptance or prejudiced inapproachability."[57] Hrdlička hoped that, by learning more about the history of the human body, people would reconsider dogmatic beliefs they clung to about humankind. He hoped that visitors would absorb some of his complex ideas about racial and individual variation, considering also the importance of race to the story of evolution. The exhibit in San Diego argued that skeletons represented discrete scientific facts. Facts were merely expounded on through art in a form that was more comprehensible to exhibit visitors. Current knowledge posited that man had clearly evolved into distinct races. Although the exact nature of these races was constantly in flux due to subtle and ill-defined environmental factors and the mixing of "pure" races, it was presumed that careful measurement and astute observation could allow scientists to represent distinct racial groups through the depiction of only a small number of supposedly representational individuals.

During the fair, the *San Diego Union* newspaper featured a column spotlighting the *Science of Man* exhibition. Written by James W. Wilkinson of the San Diego Normal School, one article introduced readers to the exhibition's importance in understanding new ideas about human evolution. The article begins, "To fully appreciate the importance of the exhibit in the Science of Man exhibition, one must bring with him a lively imagination and attempt to visualize the conditions under which primitive man must have struggled." Indeed, in order to understand the story of human history, evolution, and modern-day variation, visitors to *The Story of Man through the Ages* were expected to tie together the relatively isolated scientific facts into a more "alive" rendition, primarily through the power of imagination and fantasy. In the article, potential visitors were provided some background as to the importance of the displays about human ancestors. The final displays at the *Science of Man* exhibition lacked an overtly eugenic message, but visitors were anything but discouraged from stretching subtle arguments of the exhibition to popular eugenic ideas of the era. Along the lines of thought espoused by eugenicists, Wilkinson interpreted the exhibition as demonstrating that "the inevitable result of the spread of the doctrine of evolution will be that man will strive more and more to control the forces of nature and make them work for his lasting welfare. There will be an enlightened program favoring courageously the survival of the fittest human and the gradual development of a sturdy public opinion that will refuse to tolerate industrial and social conditions that tend toward the debasement and deterioration of the race."[58]

Museum visitors inevitably draw their own conclusions from exhibits they observe, and these conclusions are not necessarily identical with the objectives of the curatorial staff. Wilkerson, as was often true of journalists, used his own imagination to draw out a more lively story from the specimens. Despite, or perhaps partly because of, such subtle exaggerations, the exhibit proved successful, ensuring the creation of a permanent museum on the fairgrounds in Balboa Park. Over the decades that followed, other museums and anthropologists increasingly adapted similarly dramatic renditions of humanity's evolution and racial history. Despite the captivating nature of the human remains being presented, the San Diego exhibit lacked clear themes or ideas for visitors to latch onto.

While Hrdlička did recognize the importance of individual variation within races in his exhibit catalogue,[59] most visitors to the display likely would have seen how busts of racial groups were consistently segregated and thus have left the exhibition with the notion that racial typology remained the central tenet in the study of physical anthropology in the United States. Indeed, the bodies of individuals portrayed through art and bones were offered as specimens representing larger groups of populations. Visitors might have assumed that the study of ancient and recent human remains, combined with the measurements of the living, provided straightforward information about the development of humanity. Stretching these arguments to a conclusion that included particular ideas about race and eugenic "fitness" involved no great leap. Although both individual variation and the environment were put forward as significant factors shaping our bodies, race was consistently portrayed as the most important factor that defined physical characteristics. Throughout the displays, but especially in the sections where actual human remains appeared, it was evident that without direct study of human remains and the bodies of the living, no "science of man" was possible.

Both prominent public figures and museum professionals responded positively to *The Story of Man through the Ages*. George Gustov Heye, the wealthy patron and founder of the Museum of the American Indian in New York City, visited the Panama-California Exposition. Heye wrote Hrdlička afterward, "It, without doubt, is the finest showing of physical [a]nthropology that has yet been given to the public." He specifically praised the visual representation of complex ideas.[60] Heye's comments centered on the style of presentation, but he also commended the overall accessibility of complex scientific ideas about race and evolution, noting that even small children were able to learn from the exhibit. Hrdlička replied, "Though not perfect, [the displays] represent really more than has ever been attempted in these lines either in this country or abroad."[61] Sometime after viewing the exhibitions, Heye was moved to jump-start the work of the Department of Physical Anthropology within his own museum. In announcing the decision, he wrote, "It is realized that, while the creation of a Department of Physical Anthropology is a scientific need, although a complicated task, it will be our endeavor to meet all the modern demands of that science."[62] Heye ultimately experienced only limited success in founding a tradition of physical anthropology at his own museum in New York. His primary concern had always been the material culture collection—and it remained so even after being influenced by the exhibit in San Diego. Nevertheless, his decision to aggressively collect human remains after viewing the exhibition in San Diego demonstrates the remarkable, yet mostly forgotten, influence of these displays.

Walter Hough, an anthropologist at the Smithsonian, noted in an unpublished manuscript, "The exposition at San Diego is of great import to the progress of anthropology in California. There will remain in California at the close of this exposition a permanent collection relating to man that has not been excelled in this country."[63] Hough's claims, of course, were bloated with the hyperbole of world's fair enthusiasm. According to Hough, the development of the collections had implications not only for the science of anthropology but also for the progress of the entirety of culture. The science of collecting and studying skeletons, in other words, represented a major advancement in human civilization. California, at this time, was still considered by many easterners to be a dusty outpost on the edge of the continent, still young in its research and cultural production. Hough wrote, "There is being built up on the west coast a people of general culture who are appreciative and receptive of the researches of science. . . . It augurs well for the science of anthropology here that it has a public that aids the extension of its activities—a public that demands and can assimilate its result in science."[64] Hough observed that the exhibit was influencing ideas and culture in the region, and indeed, the San Diego Union enthused, "These wonderful collections are to remain as a future asset to our community and will become the nucleus for a great civic museum."[65]

Ultimately, however, some anthropologists had serious reservations about the exhibition. Scholars were in almost complete agreement that the fair brought together the most important displays surrounding race and human prehistory available in the United States, but not everyone agreed with interpretation of the materials presented, especially with regard to the development of racial groups throughout time. Hrdlička and Franz Boas frequently disagreed throughout their respective careers, though the two strong-willed individuals did, at times, manage to cooperate on specific projects. Boas dismissed Hrdlička's notion of the progress of civilization, arguing instead for a more relative perspective on the development of culture. Boas's former student Alfred Kroeber, at the University of California, sent a polite postcard to Hrdlička from the fairgrounds but chose not to engage with him as to how civilization was portrayed as evolving in "stages."[66] Such arguments depicting modern humans as evolving through various stages from savagery to civilization were coming under increasing scrutiny from anthropologists. Boas and Kroeber increasingly moved away from their own studies on human remains—though continuing to recognize its importance as a subfield—instead emphasizing work on ethnography, linguistics, and advancing theories in the field while teaching a large number of influential students. Physical anthropologists, meanwhile, were torn as to how to depict and classify the modern races—a subject that became increasingly problematic in the ensuing decades. As the issue of racial science became increasingly fraught by the middle of the century, narratives surrounding human evolution and prehistory came to play an increasingly central role in the exhibitions staged by natural history museums in the United States. Future exhibitions on the subject of racial classification and human evolution responded directly to the displays appearing in San Diego in 1915. Small reminders of the exhibition, including a number of the original busts of humankind's ancestors shown early in the displays, remain in the galleries of the San Diego Museum of Man—now heavily contextualized with panels detailing how they came to be at the museum.

. . .

Hrdlička himself was critical of the displays he created in San Diego. He noted the "untoward circumstances" preventing him from fully realizing his vision, in particular, the outbreak of World War I, though he allowed that he might be the only person who would ever notice these deficiencies.[67] His displays on physical anthropology had broken new ground, introducing visitors to new concepts and ideas. Visitors witnessed, for the first time in the United States, artistic representations and casts based on skeletal remains found in museum collections from Europe. They were also introduced to the notion that science could explain the effects of aging, individual variation, and race on the body through scientific facts and a sparse, ill-defined narrative of human evolution throughout time. Moreover, the displays related to physical anthropology were striking in their visual diversity.[68] Busts and casts of human bodies, including those of prehistoric remains from Europe, were complemented by extensive maps and charts. At the end of the exhibition, visitors encountered a large display of human remains, something rarely exhibited on this scale in museums of the era. The display of diseased or injured prehistoric human remains, in particular, was unique. Finally, visitors viewed a large series of artistic busts representing Hrdlička's principal ideas on evolution, aging, and race.

Although the displays in *The Story of Man through the Ages* at the Panama-California Exposition are largely forgotten in the annals of museum history and the history of anthropology, they provided a precedent from which later exhibitions surrounding race and prehistory drew inspiration. Museum professionals and anthropologists, in fact, wrestled with the exhibition years later when even-larger exhibitions were in the works. Not only was the exhibition influential for future museum displays, but much of the original material remained on view until the arrival of World War II—and some of the material has been reinterpreted for display today.[69] Museum curators struggled to move away from the presentation of bones as scientific objects, increasingly trying to grab hold of engaging narratives with which to teach the public. The images and art originally produced for the exhibition were even directly reproduced in future exhibitions and publications on the topic of race and prehistory.[70] The human remains seemed to even subtly drift into the background with these exhibits. Yet the San Diego exhibits made explicit to the visitor the relationship between the development of ideas expressed through art and the bone rooms behind the scenes. Exhibit curators who created new galleries on the same subjects returned to the same displays despite the fact that ideas about race and human prehistory were shifting and changing. Later exhibits, despite their turning to San Diego for guidance, were also heavily influenced by ideas stemming from human remains collections, as well as a strengthening narrative relating humankind's history.

CHAPTER 5

SCIENTIFIC RACISM
and MUSEUM REMAINS

In 1936, W. Montague Cobb, an African American physical anthropologist, published a new book about his work at the historically black school Howard University.[1] *The Laboratory of Anatomy and Physical Anthropology of Howard University, 1932–1936*, was well received by other scientists studying the human body. The biologist Raymond Pearl echoed other scholars when he described Cobb's work this way: "This account is withal so straightforward, so modest, so unselfish, and so intelligent as to win instant sympathy and admiration for its author's clear-headedness and philosophical soundness." Pearl argued that the new laboratory at Howard University should "be encouraged and supported, both from within and from without the institution."[2] Howard University in the 1930s was an important center for African American intellectuals. Cobb found support for his work at the school, which would also train the next generation of activists, lawyers, and scientists. Many of the school's students, faculty, and alumni became key participants in the civil rights movement.[3]

Cobb joined the Howard faculty in 1932. Having received training in both medicine and anthropology, he followed the model for young physical anthropologists in the era. Unlike many of his peers, however, Cobb mostly spent his career affiliated with a medical school rather than a natural history museum or anthropology department within a university. Nevertheless, Cobb eagerly sought out bone rooms across the country to find evidence to support his ideas. Later, while working to train both medical students in gross anatomy and anthropology students in comparative anatomy and evolution, he helped amass over seven hundred skeletons for collections at Howard. Cobb acquired many skeletons from the cadavers dissected in his anatomy laboratory.[4] Steadily gaining respect in the anthropological community in the face of both subtle and overt racism, Cobb was ultimately elected president of the American Association of Physical Anthropologists in 1958. Throughout his career, Cobb managed to leverage his standing in the broader scientific community to promote racial equality.[5] Cobb also served as president of the National Association for the Advancement of Colored People (NAACP) between 1976 and 1982.[6] Having received his training in an era when most African Americans were denied equal access to faculty appointments at most universities, for a time he was the only prominent black scholar in physical anthropology. His influence in the anthropological and medical community was extraordinarily vast, having authored over eleven hundred publications and trained more than six thousand students in anatomy.[7] Cobb was especially prolific in writing newspaper and magazine editorials condemning unequal medical treatment for African Americans.

In a 1936 monograph describing facilities at Howard, Cobb tied together anatomy and physical anthropology as disciplines, detailing the medical school curriculum alongside a program for teaching physical anthropology and human evolution. Aleš Hrdlička, who had continued to argue in favor of general medical education for physical anthropologists, applauded his efforts, offering in a letter that bore a decidedly paternalistic tone, "You have a rare chance for the development of a department which will be a model for Universities for colored people elsewhere in the country." Hrdlička continued by articulating his opinion of the value in bridging disciplines in the modern university: "And you have done wisely in associating anatomy with physical anthropology, for the latter is in a large measure merely advanced comparative human anatomy, and aide [sic] the student to become acquainted with human variations, with which he will everywhere be confronted after he leaves college."[8] In Hrdlička's view, in other words, educated individuals should be able to understand race as a concept on a physical level.

Hrdlička's reaction to Cobb is striking, as only a little over a decade earlier, as he baldly declared his opinion that "black people represent a mental potentiality, say, only 80 percent of the average white people."[9] While Hrdlička's opinions on the subject of racial intelligence shifted somewhat over the course of the ensuing decade, it is also possible this comment rested on the belief that individuals within any racial group were capable of high scholastic achievement. Regardless of the origins of Hrdlička's remark, the conclusions drawn by the two scholars should not be thought to be identical. Whereas Cobb's research, based heavily on his work with collections of human remains, pointed to physical similarities between races, Hrdlička had built a career on differentiating and classifying them. Others scholars, in the face of criticism, extended the claim that the entire history of human achievement could be tied directly to features of anatomy. The primary source of evidence, central to these studies, was in the collections of human remains stored in bone rooms that had grown since the Civil War.

Scholars largely point to the era between 1920 and 1945 as the period when scientific racism came under increased scrutiny in the United States. The historian Richard King argues that intellectuals and scientific elites during this period largely discredited the concepts of race, racial difference, and racial hierarchy.[10] Eugenics and scientific racism became increasingly linked to the destructive politics of the National Socialist Party in Germany, despite the divergent landscapes of science in Europe and the United States.[11] Journalists, politicians, and some scientists linked the scientific racism of scholars in the United States directly to the dark philosophies of the Nazis, pushing the race concept further out of the mainstream in many scientific fields.[12] Even before these extreme expressions of scientific racism, however, scholars in the United States who were linked to the practice of collecting and studying race using human remains began questioning ideas associated with eugenics. Students of physical anthropology, including Franz Boas and W. Montague Cobb, played an important role in the demise of racial science, and yet the transition from supporting racial classification to arguing against it among those who studied human bodies was never complete or entirely decisive. Cobb based his scholarship on directly examining human remains collections. He then broke down strict forms of racial classification, which he viewed as both inaccurate and inextricably linked to scientific racism.[13] Scholars of medicine and physical anthropology wrestled with the meaning of these transitions in light of a lengthy history of collecting human remains for museum and university collections. These tensions were on display in both publications and museum exhibitions. Many sought to hold onto older forms of racial classification, while others gradually shifted their attention to other questions entirely. While scholars such as Cobb promoted the study of physical anthropology through the collection and display of human remains, others, like Boas and his students, shifted away from measuring bones and bodies and toward a more theoretical study of the concept of culture.[14] Boasian anthropology, promoting cultural relativism, gradually became the dominant conviction in the field throughout the middle portion of the twentieth century in the United States, though other anthropologists, such as Carleton S. Coon at Harvard University, did write popular books promoting ideas of racial hierarchy well into the 1960s.[15] Human remains collections both informed these studies and were influenced by the ensuing discourse surrounding the validity of their claims.

Human prehistory was a component of studies of human remains long before the 1920s. The shift away from the race-centered study of bone collections to studies based more frequently in questions about prehistory and population came about gradually and sporadically. Once these shifts started taking place, however, the work of those who were focused primarily on racial classification schemes and the studies responsible for the creation of bone rooms to start with began to be viewed as outside the mainstream of American anthropological thought. This slow decline of racial classification as a scientific and intellectual concept was reflected by scholars who were concerned with human remains, particularly those who were previously concerned with collecting human remains to justify racial classification schemes. Whereas displays on eugenics were prevalent at fairs and museums in the early years of the twentieth century, such displays were removed from plans for future exhibitions, replaced by a growing emphasis on human history.

The shift away from racial classification created a tension in the long-standing practice of collecting and interpreting museum collections of human remains. In the early 1920s, museums were organizing human remains by race or geographic region, rather than by age. By the late 1930s and early 1940s, though, museums began to shift emphasis away from race as the main defining category for human remains collections. Visitors gradually encountered exhibitions that mirrored the manner in which anthropologists conceived of the remains, gradually emphasizing human history alongside galleries on race.

AMERICAN MUSEUM OF NATURAL HISTORY

Sometime in the early 1920s, curator Clark Wissler penned a brief description of the physical anthropology collections at the American Museum of Natural History (AMNH) in New York. Wissler's manuscript offers a glimpse as to how collections were brought together and studied during this period at one of the largest museums in the nation. The document begins by noting that the collections of the AMNH contained over twenty-one hundred human crania. The emphasis on skulls was the result of several influencing factors. First, building on traditions ranging from phrenology to medicine and anthropology, scientists continued to believe that the human skull held clues to answering the most important questions about race and human history.[16] Skulls were also more durable and easily transported than other body parts; smaller, more fragile bones, although still prized for museum collections, were often lost, broken, or sometimes even simply discarded. Two large collections made up a sizable portion of the total number of crania at the museum; Eskimo crania accounted for 350 items in the total collection, and nearly another 250 were brought to New York from the American Southwest.

Apart from the human skulls of North America, the largest number at the AMNH originated from South America, with an additional 600 crania from Bolivia and Peru and another 350 from Mexico. The museum featured a smattering of remains from other regions around the world, listing "2 Pygmies, 3 Australians, 2 Japanese, and 1 New Zealander."[17] Although race had been the defining factor driving the actual collecting of human remains for the previous fifty years, remains were sometimes classified internally by region or geography and were increasingly so with the passing decades and changing scientific and intellectual milieu.

Although Hrdlička visited the AMNH to study Eskimo skulls, Wissler notes that, at that time, the vast majority of the cranial collections in New York remained unstudied.¹⁸ Wissler's description did not identify any mummified remains from Egypt, though collectors from the United States and Europe were still actively buying, and subsequently donating, mummies. Though the manuscript hints at the prehistoric nature of some remains, it briefly organizes them by region or race rather than making any sort of claim about the age of the specimens. Within a decade of the circulation of the memorandum, however, the emphasis of many major museums in describing their human remains collections changed. Race was the major qualifier for human remains early in the 1920s, but reinvigorated interest in human prehistory reshaped how remains were conceptualized in museum collections in the United States in later years. Mirroring the emphasis on race found in the AMNH's description of its own collections was work in the field of anthropometry.

ANTHROPOMETRY AND MEASURING RACE

In 1920, Hrdlička published a detailed volume providing instruction on the subject of anthropometry, which he defined as "the conventional art or system of measuring the human body and its parts."¹⁹ The term *anthropometry* was more often used in reference to the practice of measuring the living, but in his book, Hrdlička also included instructions for measuring bones. Above all else, Hrdlička privileged the standardization of measurements and thus the presumed elimination of personal bias from the science.²⁰ According to Hrdlička, measurements of the human body were useful for industry, art, the military, medicine, "detection of bodily defects," the identification of criminals, eugenics, and general scientific investigation.²¹ The fact that bodily measurements were perceived as being so useful to a variety of fields created the drawback that Hrdlička rued: that of having vastly different systems for measuring and interpreting the body. At heart, Hrdlička indicated, anthropometry was about the development of complex systems for understanding human physical appearance and behavior. For Hrdlička, as well as for numerous other scholars at this moment, the shape of the skull, though not as indicative of human behavior as earlier phrenologists or criminologists believed, could teach scientists about certain aspects of intelligence and behavior. Evidence collected from an untold number of expeditions—where indigenous peoples were measured and remeasured—filled anthropologists' notebooks in both museums and universities. Complementing the measurements collected from outside the territory of the United States were those collected from indigenous peoples who visited world's fairs or major cities, where anthropologists were eagerly waiting to measure their bodies. Despite the thousands of measurements collected from the living throughout the late nineteenth and early twentieth centuries, scholars continued to believe that the most accurate and lasting baseline for these types of calculations would stem from work with the human remains stored in bone rooms.

The system that became standard in the field of anthropometry valued consistency, precision, and simplicity. Researchers needed large amounts of data from every part of the globe. In order to acquire the data, groups of trained practitioners could gather consistent measurements and publish them for future research. The First World War, Hrdlička repeatedly noted throughout his career, created a number of opportunities to collect general measurements of human bodies, including thousands of measurements of soldiers, which he presumed would be useful in future studies. While these opportunities to collect on a truly massive scale were not fully realized at the time, the experience was thought to provide lessons for future conflicts. Compared to the practice of collecting human remains from gravesites that were sometimes physically guarded and generally regarded to be sacred, measurements of the living were collected with ease.

When selecting living subjects or skeletons to measure, Hrdlička argued that the most significant factor to consider was simple—race. He wrote, "In the study of any human group the value of the data—all other things being equal—will be directly proportionate to the purity of the group."[22] The "purity" of any racial group, he elaborated, could only be determined through consistent measurements and interviews with the living to determine familial heritage. This information, it was concluded, would help determine the racial heritage of skeletons being measured in museum collections. Age, sex, and medical history were also recorded, in addition to information about the subject's social status, occupation, and "environmental distinctions."[23]

Hrdlička classified the mixture of races, to his audience interested in anthropometry, as occurring in two different forms: that which occurred between tribes but within larger racial groups and that which occurred with the mixture of blood between differing racial groups. Hrdlička notes that evidence of admixture between tribes—but within races—could hardly be determined without associated family history. Mixture between the major racial groups (considered at the time to be white, black, and yellow-brown) was thought to be more easily determined through measurements and observations alone.[24] The desire to collect anthropometric measurements reflected the task of clarifying these determinations.

Scholars who worked with skeletons, as opposed to the living, did gain several advantages. Hrdlička described the study of skeletal remains as "a particularly attractive field, for we deal here with specimens that are not masked by other tissues, that can be handled cleanly and easily, and that are mostly and completely at our disposal for reference or additional observation."[25] Hrdlička sometimes disparaged the anthropological obsession with the human skull, but he recognized and even advocated for the utility of the cranium for understanding the entire human body.[26] Unlike a living subject, the bones always refuse to answer direct questions regarding such things as family history, and thus information understood as providing clues to racial history was lost. While some skeletal material arrived at museums with archaeological information that provided some information about ancestry, many skeletons arrived with only limited information about their acquisition. Nevertheless, Hrdlička argued that even without a family history or contextual archaeological information, the race of the subject might still be determined through measurement and comparison with existing data: "Recognition of distinct racial types in a collection, demands especially careful procedure. The skull of a typical White, a typical Negro, a typical Eskimo, or a typical American Indian, may be readily and reliably identified, wherever found by the expert student; and in a smaller measure this is also true of some other parts of the skeleton."[27] Just as with the bodies of the living, Hrdlička noted that the skeletons of individuals with mixed ancestry were harder to identify: "But when it comes to a recognition of crania or bones of mixed-bloods, or of closely related racial types, we face considerable uncertainties. The safest rule in all cases is for the observer to set aside from his series any skull or skeleton concerning the anthropological identity of which he is in serious doubt. He will bear in mind, of course, that among all peoples there exists in every feature a wide range of normal variation."[28] Despite the apparent certainty of racial groups, scholars were forced to recognize the fact that human remains across all racial groups reflected a wide variety of features. Coupled with concerns about increasing mixture between racial groups, some indications were already showing that the basis of systems of racial classification had serious flaws.

Instructions given to scholars interested in anthropometry in the early 1920s focused primarily on racial classification. In writings on the subject, Hrdlička focused largely on measurements of the living. Skeletal remains, on the other hand, held certain clear advantages for the anthropologist. Consistent and complete sets of data, reflecting the diverse range of ages and genders within what were considered "pure" and "mixed" racial groups were actively sought. Bodily measurements of the living were seen to be complementary, providing information about the bodies of the dead. Just as with Wissler's description of the skeletal collections in New York, Hrdlička's guide to anthropometric measurements emphasizes race over the age of the human remains being measured. Measurements of the living were to be collected, stored, and published alongside collections of human remains. The measurements of the living, as well as the remains of the dead, were thought to speak to the understanding of humankind. Instead of addressing questions about human history and ancestry, however, these types of materials, up to the early 1930s, were primarily understood as contributing to knowledge of race. What was at one point a seemingly singular emphasis on race, however, started to gradually change among scholars collecting bones for museums in the United States.

AMERICAN SCHOOL IN FRANCE FOR PREHISTORIC STUDIES

In July 1921, several American students began work at the newly created American School in France for Prehistoric Studies (ASFPS), later the American School of Prehistoric Research in Europe.[29] The newly created school served several functions. The first was to find new prehistoric specimens in Europe, including archaeological material such as stone implements and animal bones. Once new specimens were found, the group investigated their significance by comparing them to available collections at museums in the region. Following excavations, discoveries were split between museums in France and in the United States. The study of prehistory was thus intimately linked to the nationalistic politics that marked Europe in the late nineteenth and early twentieth centuries. Museums in

Germany, France, and Britain competed for material culture artifacts, in addition to human remains, racing to build the largest and most encyclopedic record of man and the natural world. The British Museum and the Musée de l'Homme (Museum of Man), like the Smithsonian Institution or the Field Museum of Natural History, rapidly collected both human remains and archaeological antiquities.[30]

Although the rise of archaeology and anthropology followed a particular, complex history, a common thread was the sporadic participation of American museums and their staffs in the collecting of prehistoric artifacts in Europe. Europeans had been collecting ancient artifacts for centuries, but the study of ancient humans on the continent became increasingly professionalized and wrapped up in political projects of nation building in the nineteenth century.[31] Although the rise of ancient history was wrapped up in abstract, nationalistic competition between countries, museums, scientists, students, and explorers from the United States were often invited to participate in excavations. Students like those in the American School of Prehistoric Studies paid tuition and were eager to help collect prehistoric artifacts, shipping them to museums within the European nation in which they were found and occasionally sending "duplicate" objects back to museums in the United States.

In the twilight of the Victorian era, European prehistory became an era of interest and exploration for a modest number of scholars in the United States. Henry Field, who later played a major role in the development and display of physical anthropology collections in the United States, claimed to have found his first prehistoric artifact in Europe at age six, after his mother was remarried to a man with a two-thousand-acre estate in the countryside of Leicestershire, England.[32] Another young scholar, George Grant MacCurdy, trained at Harvard before spending much of his career at Yale—with stints in Paris, Berlin, and Vienna rekindling his commitment to studying prehistory.[33] The rise of prehistoric studies in Europe influenced similar studies in the United States, including those linked directly to the study and display of human remains. The opening of the school helped usher in an expanded interest in preliterate history outside of the Greco-European histories that defined the continent of Europe for most Americans.[34] Although the gravity of the focus of physical anthropology in the United States gradually shifted to deeper studies of human ancestors, mainly in Africa and Asia, the study of prehistory in Europe was, for a time, critically important to the history of human remains collections in the United States.

The ASFPS viewed itself as fighting both looters and Mother Nature, both of which undermined efforts to preserve evidence of prehistoric humans. An article summarizing the work of the school reads, "museums are the stations in which specially prepared sections of the relic-bearing deposits are protected from ruthless hands as well as from the elements, and will ever remain to tell the story of how man lived and how long he lived before the dawn of history."[35] The school also proclaimed itself to be serving a new function in American archaeology, described by the director of the program in his comments on program activities: "They were undertaken in the spirit of the pioneer, who has no precedents to break and none to observe."[36] The ASFPS, though certainly not the first official effort to obtain prehistoric material for museums in the United States, signals something of an official movement toward prehistory in American archaeology.

In the summer of 1923, Aleš Hrdlička served as the school's director, though he was hesitant to take on the task of organizing and maintaining the field school. Hrdlička, as might be expected, instructed the students to read background literature on both general prehistory and his work on ancient skeletal remains.[37] In addition to expanding the school's interest into the realm of human remains, Hrdlička also hoped to expand the school geographically to other parts of Europe. This decision, however, was unpopular among other scholars involved with the school. Charles Peabody, the curator of European archaeology at the Peabody Museum at Harvard, was led to comment on Hrdlička's plans, "I am sorry he has departed so far from the ideas of those of us who founded the School. It is hardly the 'School in France.'"[38] Hrdlička's intentions to expand geographically were pedagogical; however, in addition to having the students read about the use of skeletal remains in the understanding of prehistory, Hrdlička brought the students to a field site where Neanderthal remains had been discovered.[39] Hrdlička argued, and the committee that supported the school eventually agreed, that "the School ought to ... give the American students the very best possible," which, Hrdlička explained, "should include an initial firsthand knowledge of the most important site and discoveries of Early Man in Western and Central Europe."[40]

It would be inaccurate to describe the collecting of prehistoric remains in Europe as totally separate from the project of nation building in the United States, but this history was more directly tied into nationalistic competition between the museums of Europe. The professionalization of prehistory in Europe, however, influenced museums in the United States and encouraged scholars to look for clues about humanity in the bones of the dead from across all of history. Although studies in prehistory were not without flaw, having been tied to complex projects of nationalism and imperialism, they stole attention away from the racial classification theories built on the study of the remains of the ancient dead and the measurement of the living.

HUMAN REMAINS AND ANCIENT HISTORY

As a handful of scholars interested in collecting, studying, and displaying human remains began to influence the study of prehistory, the practice spread into other branches of archaeology—including the ancient history of early literate and classical societies. In a letter to the revered Egyptologist James H. Breasted of the Oriental Institute at the University of Chicago, Hrdlička expressed his concern about the fate of remains found on archaeological expeditions that were focused mainly on ancient history. Hrdlička stated that he could not help but think of the skeletal material that the Oriental Institute would inevitably uncover while conducting archaeological investigations. He articulated his hope that the archaeological work of the institute would be "attended by a saving, as far as possible, of the precious skeletal material that may be discovered"; he continued, "the anthropology of the Near East can never be well understood without a study of ample skeletal remains of the people."[41] Just like the world war and various international expositions, new archaeological expeditions to the Near East—including several led by Breasted and the University of Chicago—provided opportunities to collect skeletal remains. The study of ancient history, to be sure, was tied into the construction of nationalistic projects just as the study of prehistory was, but the growing emphasis on human remains within the study of ancient world cultures also pointed to burgeoning questions about human history more generally. Race as a subject certainly did not escape the equation, as scholars who studied ancient skeletons—stretching back to Samuel George Morton's work in the mid-nineteenth century—continually turned to ancient remains for clues about the supposed solidification of racial characteristics. Furthermore, observers in the United States often tied ancient remains, like those from ancient Egypt, to particular biblical narratives, thus increasing their popular appeal. As the fields of physical anthropology and archaeology continued to professionalize in the United States, ancient remains displayed at museums and fairs were less often tied to particular stories from the Judeo-Christian tradition.

Museums in the United States during this period were especially active in collecting classical or Old World archaeological material. Professional collecting of material from present-day sites around ancient Babylonia, Egypt, Greece, and Rome was fueled by an existing familiarity and fascination with these regions. Professional archaeological associations allowed museums in both the United States and Europe to fund expeditions and in return to receive a portion of the materials discovered. Outpacing the professional collecting of classical archaeological materials were the private donations of wealthy patrons. When not purchasing objects on their trips around the world, wealthy patrons of museums in both the United States and Europe frequently funded archaeological expeditions, hoping to fill their favorite museums with priceless artifacts from various parts of the globe. These material donations frequently included mummies from Egypt, which became popular attractions for museums that displayed them. Many people in the United States were familiar with ancient Egypt due to its centrality in biblical narratives, and the opportunity to view the preserved body of an ancient Egyptian proved to be alluring. Museums, in turn, were encouraged to display donated mummies due to both their widespread appeal and the expectation that remains of this kind would be on display in museums of science, history, and natural history. The practice became so ubiquitous, in fact, that one museum curator was later prompted to reflect that no self-respecting museum was without a mummy from ancient Egypt.[42] Although these remains may have been exhibited and observed with an Orientalist, racialized appeal, their primary draw in the early twentieth century was their direct connection to the ancient history that was more often found in Judeo-Christian narratives.

ROLAND B. DIXON AND RACIAL HISTORY

Despite growing interest in prehistoric archaeology in the United States, race continued to be dominant in physical anthropology of the first quarter of the twentieth century. Roland B. Dixon, an anthropologist at Harvard, published many of his ideas in a controversial volume titled *The Racial History of Man*, in 1923.[43] The reception of the book was lukewarm, at best. Dixon wrote to Hrdlička, "I know that I can hardly expect that it will have a very favorable reception, for you will, I am sure, regard the method as wholly indefensible. I beg, however, that you will regard it as an honest effort to try to bring together in one single field of view, a terribly complex subject."[44] Hrdlička replied with a blend of American metaphors: "As to the book, all that I can say is that if you have spilled the milk you will have to take your medicine."[45] Following the publication of *The Racial History of Man* and the publication of a major catalogue of the Smithsonian's human crania collections, Dixon and Hrdlička traded blows in a series of letters and harsh reviews.[46] The two squabbled over the correct approach to measuring skeletons. While seemingly trifling points, the debate held major implications for the two men with regard to the way racial categories were drawn. If they could not agree on the manner in which crania of different races were compared, for example, the influence of their respective work would be hampered. Not only did they argue over methodological details such as measurements, but the two also argued over the nature of Hrdlička's control over the Smithsonian's collection of remains. Dixon accused the Smithsonian of "withholding from students important data which were actually in [Hrdlička's] hands."[47] In responding to Dixon's nasty reviews and letters, Hrdlička wrote curtly, "I know that revenge is sweet, but you are an inordinately ungrateful and greedy lot, all of you. Also, most unreasonable."[48] The race to acquire remains from around the world—a process that had previously manifested itself in a seemingly friendly, scholarly competition—had become a rivalry over bone rooms in museums across the United States that suddenly took a hostile, and quite personal, turn.

Disagreements between the two scientists hinged on two major points. Intellectually, they argued over which skeletal measurements were most useful in differentiating various populations. Certain measurements, each of the men argued, were more useful and stable in the comparative study of race, while others were virtually useless due to fluctuations or lack of stability within groups; these disagreements point to some of the flaws that scholars later used to critique the overall methodology of physical anthropologists. Further, Dixon argued that Hrdlička mishandled his position as the curator of the nation's largest collection of skeletal material. Dixon wrote to Hrdlička, "You are, in a relation to the collections in your charge, not a private individual. You are a trustee for scientists everywhere. You have no right to follow your individual opinions in regard to what you shall publish." Dixon continued by arguing that the data Hrdlička must make available to other scholars should be as complete as possible: "It is your duty to afford to others the most complete information possible, when you publish an official catalog of the national collections."[49] That key measurements were missing from the Smithsonian catalogue rendered the volume entirely useless as a scholarly resource. Although other scholars hesitated to say so directly in correspondence and publications, the tone of Dixon's comments echoes the widespread unrest over Hrdlička's restrictive management of Smithsonian collections. This type of unrest cut across numerous institutions and professional relationships, manifesting itself in the form of personal rivalries like Hrdlička and Dixon's, fueled in part by the sentiment of competition between different institutions over the collection, display, and interpretation of human remains. Underlying the idealistic arguments for academic freedom or access to collections were the real sentiments of institutional and scholarly competition over human remains. These rivalries had certainly emerged in Peru years earlier when Yale attempted to claim scientific rights over natural history and archaeological collections—including human remains—and they arose again as Harvard and the Smithsonian hashed out methods for studying skeletons. Throughout all of these debates, collections of skeletal remains were conceptualized as scientific objects, rather than the bodies of the dead, a fact that seemed to barely be referenced at all.

Some weeks later, Hrdlička wrote to Dixon asking that he explain which measurements, specifically, he hoped to see included in the Smithsonian's catalogue. Dixon sent a letter, now in a much calmer and more measured tone, arguing that the Smithsonian's official catalogue of crania should include several new measurements. Dixon was sensitive to the notion that adding space for new tables containing measurements might be a challenge. Dixon indicated his opinion that space for these new measurements might be acquired by relegating to footnotes existing space in the catalogue dedicated to listing how skulls were obtained. Details about deformities, too, could be abbreviated to save space. Dixon clung to the belief that cranial measurements would be useful for the comparative study of race.[50] For him, the exact archaeological provenance and the nature of various deformities were simply afterthoughts for the development of racial classification theories. Hrdlička eventually agreed to include several of these requested measurements in future catalogues but reminded Dixon that he could simply write to ask for more specific details about each cranium.[51] Despite these eventual compromises, the perception remained that Hrdlička was aggressive and highly guarded in his curation of the Smithsonian's collections. Leaders at Harvard, as elsewhere, periodically indicated their desire for greater access to these collections in order to advance competing ideas about race and history.

Not only did *The Racial History of Man* fuel arguments between Hrdlička and Dixon about curatorial control, but it led other scholars to critique racial theory using evidence in the book. Franz Boas began his critique of the book by detailing the role of standardized measurements in the field of physical anthropology and the study of the ancient history of humankind. Whereas previous scholars relied mainly on observation of physical characteristics, Boas argued that modern scholars attempted to quantify these differences through careful, scientific measurement. Boas wrote, "Professor Dixon's attempt to unravel the racial history of man runs counter to this whole development." Dixon's book argues that the physical features of the eight races of humankind have remained relatively stable over the course of time, an argument that many scholars of the period were rejecting increasingly emphatically. According to Dixon, humans were thus immune from environmental influences—an argument that scholars like Boas tore apart based on their understanding of evolutionary theory of the period. Though Boas did argue, "It is, of course, true that the human races have intermarried to such an extent that the attempt to find a pure race anywhere is futile," he continued that, "notwithstanding this fact, we ought not to overlook the similarity of the phenomenon to the analogous variability of plants and animals which occur over extended areas." Boas was also attuned to subtle claims of racial superiority made in Dixon's argument. Dixon argued that the history of human achievement might be understood singularly through the study of anatomical form. Boas argued that, just as was true with human culture, the human form had changed gradually over time. The emerging evidence, according to Boas and many of his later followers, suggested that races could no longer be conceptualized as belonging to unchanging racial categories. Boas therefore replied to Dixon's competing notion of racial stability directly, arguing, "If it were valid, then at different periods it would justify entirely different views." In Boas's view, cultures ebbed and flowed in their relative strength and achievement, and not along lines of strict morphological characterization. Boas further implied that scholars like Dixon largely changed their tune depending on the context, their arguments shifting from discussions of the overall racial superiority of whites to critical responses to claims of racial superiority emerging in Germany.⁵² The concept that Dixon proposed meshed well with the popular eugenic theories of the moment, but scholars like Boas and Hrdlička were seemingly too busy critiquing other aspects of the volume to take note of that fact. Instead, Boas critiqued the notion of racial stability, while Hrdlička derided Dixon's methodological approach to the study of human bones.

The concept of "racial history" was not new when Dixon utilized the term for his book. Some of the backlash to the book, in fact, arose from Dixon's willingness to conflate the various terminologies, theories, and methodologies of prehistoric archaeology with those of the study of race. Many scholars clearly believed that these arenas were rightly starting to move along differing courses by this time. Dixon's ideas, in fact, were in some sense closer to ideas postulated earlier in the century. Thomas Wilson, a Smithsonian curator who specialized in prehistoric archaeology, described the early germination of racial history in an undated and unpublished manuscript. Wilson, who died in 1902 (placing the origin of the manuscript at least twenty years prior to the publication of Dixon's *The Racial History of Man*), described the relationship between the study of race and prehistory in simplistic detail: "Any comprehensive study of the races of man should begin with his origin, if only to give a resume of the theories [*sic*] advances." Wilson, in his earlier work, then turned to a series of questions that scholars of Dixon's generation would have found outdated, including the old polygenesis and monogenesis debates that were more commonly associated with the study of race in the middle of the nineteenth century.⁵³ As the majority of scholars pushed back against the overall concept of racial history, older methodological approaches to the study of race that utilized the discipline of prehistoric archaeology as a major tool were still in existence. Despite the barrage of critique, echoes of the racial classification theories of the turn of the century were seemingly revisited in the pages of significant works published decades into the twentieth century. Although these debates may have centered on the intellectual interpretation of evidence drawn from bone rooms regarding racial history, they were also firmly ensconced in the institutional competition continually appearing between museums in the United States during this period.

EARLY GROWTH OF PHYSICAL ANTHROPOLOGY IN UNIVERSITIES

By the mid-1920s, physical anthropology was growing in universities in the United States. Fay Cooper-Cole, the founder of the anthropology department at the University of Chicago, reported to Hrdlička in 1926 that students of physical anthropology were making particular strides within his department. "The interest here is keen," wrote Cole, who viewed Hrdlička as a leader in the field. Cole hoped Hrdlička might visit the campus of the University of Chicago, explaining, "we are anxious to have our students meet the leaders in American anthropology."[54] Cole had spent time working at the Field Museum of Natural History before leaving for a job at Northwestern University, followed by a permanent appointment at the University of Chicago. While Cole was a professor at Chicago, his stature in American anthropology grew. By 1925, he was asked to testify as an expert witness in the notorious Scopes "Monkey Trial." Cole, knowledgeable in both religion and human evolution, even developed a cordial relationship with both attorneys from the famous trial concerning the teaching of evolution in public schools, Clarence Darrow and William Jennings Bryan. Like other academics teaching physical anthropology within departments of anthropology, Cole taught students at the university before sending them to complete their training through studying human remains firsthand in museum collections.[55] Although the growth of physical anthropology in American universities was apparent, it was stunted by some lingering confusion regarding qualifications for the field. Physical anthropologists of the era were trained disparately, in either medical schools or departments of anthropology. Just as debates were emerging regarding the standardization of measurements of skeletal remains, scholars were also arguing over what served as qualification for future scholars conducting research in the field. The struggle to determine how best to train future generations of physical anthropologists worked to hinder the early development of professional organizations and societies. The early growth of physical anthropology in universities like Chicago depended in no small part on the availability of human remains collections for study. For a time, faculty and graduate students from the University of Chicago relied heavily on the skeletal remains in the bone room at the Field Museum for their research.

Chicago was not only a place for scholarly discourse interpreting human remains. Visitors of all kinds encountered bones on display in the city's museum. For many, viewing human bodies on exhibit proved a halting and confusing experience, unlike anything they had previously encountered. In the early 1920s, when the reporter Ben Hecht published a collection of *Chicago Daily News* columns as a book, *A Thousand and One Afternoons in Chicago*, it featured the story of a two-bit criminal and "veteran con man" known by the nickname Dapper Pete. In one scene, Dapper Pete describes his experiences evading police in plain sight by visiting major landmarks in the city. Describing his time wandering through the unfamiliar museum exhibits, he remembers, "Well, then I spent three days in the Field Museum, eyeing the exhibits. Can you beat it? I walk around and walk around rubbering at mummies and bones and—well, I ain't kidding, but they were among the three most interesting days I ever put in. And I felt pretty good, too, knowin' that no copper could be thinking of Dapper Pete as being in the museums."[56]

RACE AND RUNNERS

The one-hundred-meter dash in the 1932 Olympics witnessed a dramatic finish. A pair of runners, Ralph Metcalf and Eddie Tolan, sprinted to the front of the group and lunged toward the tape, crossing the finish line at virtually the same instant. The pair of runners, both of them African American, won gold in both the one-hundred- and two-hundred-meter dashes, setting new Olympic records in the events. Ed Gordon, another African American athlete, won the gold medal in the broad jump. For some white spectators, it seemed that as soon as African Americans were allowed to participate in Olympic events, they began to dominate the competition. In the early 1930s, this seemed especially true in track and field, where a small number of athletes excelled in sprinting and long-jumping competitions. Witnessing black competitors dominate the global competition caused certain spectators to ponder whether African Americans possessed some sort of unique physical advantage. Although the achievements of African Americans in athletics were seemingly unrelated to intellectual and cultural trends in physical anthropology, they reignited popular debates about racial difference in both Europe and the United States.

W. Montague Cobb, the African American physical anthropologist at Howard University, responded directly to the preponderance of these ideas in a popular and influential article titled "Race and Runners," first appearing in the *Journal of Health and Physical Education* in 1936. In the article, more or less aimed at a popular audience, Cobb argued that these kinds of questions were by no means new. In the 1910s, Finnish dominance over other nationalities in long-distance running caused spectators to ask questions about the nature of Finnish physique and culture. Scholars of the period asked similar questions regarding whether the physical or cultural characteristics of Finns helped them to compete in a particular sport. In order to address questions of racial difference and athletic capability, Cobb examined several athletes, including the famed sprinter Jesse Owens. Cobb came to the conclusion that while these *individual* athletes certainly possessed unique physical attributes, allowing them to run faster and jump higher than the average person could, these characteristics were not *racially* unique. In fact, an examination of Jesse Owens's calf muscles revealed numerous characteristics that Cobb identified as Caucasoid rather than Negroid.

Cobb concluded, "The physiques of champion Negro and white sprinters in general and of Jesse Owens in particular reveal nothing to indicate that Negroid physical characters are anatomically concerned with the present dominance of Negro athletes in national competition in the short dashes and the broad jump." In fact, he continued, "There is not a single physical characteristic which all the Negro stars in question have in common which would definitely identify them as Negroes."[57] Despite the seemingly trivial nature of the question—whether racial characteristics influenced the ability of certain individuals to excel at sport—the outcome of the debate was telling, as it opposed not only popular belief but also the lingering tendency toward racial classification in the sciences. Cobb marshaled his extensive experience studying human anatomy—especially his experience working with human remains—to use the debate as an opportunity to refute the validity of racial categories.

THE PHYSICAL CONSTITUTION OF THE AMERICAN NEGRO

Two years following Cobb's publication on the physical characteristics of African American athletes, he wrote a lengthy article exploring current understandings of the physical makeup of people of African descent more generally. Cobb's study was based both on anthropometric work with living populations of African Americans and on extensive work with skeletal remains in museums and universities. In the opening lines of the article, Cobb cites a survey of human remains collections at American institutions, revealing that "the bulk of such material consists of skeletal remains, most of which are American Indian. But 5 per cent are American Negro."[58] Early on in the article, Cobb cites Hrdlička, who had previously argued that the understanding of the physical makeup of the American Negro was limited, at best. A comprehensive survey of anthropometric research on African Americans revealed a mere six studies at the time of Cobb's writing. Further justifying his work, Cobb notes, "Existing social conditions excite a particular interest in the nature and significance of the distinguishing features of the American Negro."[59] The nature of Cobb's writings, which at times paired social dynamics with personal conclusions about race, underscores some of his motivations. Nevertheless, conclusions were tied to careful and rigorous measurement and quantification of data, and his colleagues in the physical anthropological community clearly respected that Cobb's work was based on the supposedly "hard" evidence of actual collections of human remains and detailed measurements of the living. Within the small community of serious scholars routinely working with these collections, it would have been clear that Cobb was basing his argument on a deep and extensive record of close examination of remains of African Americans. By this point, Cobb had already supervised the dissection and removal of skeletons from countless cadavers. Although certain scholars took issue with his argument, it was virtually impossible to take issue with the depth of his experience.

In summarizing the work of scholars who examined human remains, Cobb points to numerous disagreements over the nature of the skeletons of individuals of African descent. Part of the problem was that only a limited number of remains were available for scholars to examine; when compared to the larger collections of American

Indians, the number of remains of individuals of African descent appeared paltry. While the collection at Case Western Reserve University had 800 skeletons of "American Negroes," and the collection at Washington University held nearly 550 complete skeletons of African Americans, collections still offered comparatively small samples sizes, especially when put in the context of being charged with answering questions about such vast topics as race or human evolution. Scholars working with these collections revealed different measurements than did Hrdlička, who himself had surveyed a collection of 56 "full blood" black skulls and 122 skulls from West, East, and South Africa.[60] When skeletal collections were successfully compared, however, the supposedly strict lines of racial classification were increasingly blurred, rather than coming into greater focus.

Hrdlička, in his earlier study of the collections of skeletons of African descent, had argued that the remains showed a preponderance of a premature fusion of the sagittal suture of the skull. Scholars working with the larger collections at Washington University and Western Reserve University failed to see a similar trend, and Cobb countered, "the incidence of premature union of the sagittal suture seems unwarranted."[61] While Hrdlička's study of a particular part of the development of the skull did not appear to point directly to a racist conclusion of inferiority, it did continue the long-held scientific trend of supporting claims of racial difference. These subtle differences, it was assumed, differentiated various races enough to make racial categories possible to detect with rigorous and repeated measurement. As in Cobb's smaller study of living African American athletes, he continued to emphasize the commonalities among races when examining collections of human remains. Though certainly aware of particular racial differences between bodies, Cobb's work deemphasized and critiqued the prevailing notions of racial classification and was supported through the direct examination of the human body. In the conclusion of his major survey of African American physical characteristics, Cobb remarked, "The evidence now available shows clearly that racial characters are largely variations of form which have no distinct functional survival value in modern civilization."[62] This was a bold statement for the era in which it appeared. Cobb concluded that racial differences might indeed exist and continue to be classified, yet he felt that these differences held no significance in modern societies.[63] Although it was certainly not clear at the time, Cobb's work served as something of a bellwether for trends to come, as new generations of physical anthropologists worked to further break down heretofore dominant ideas about the existence of particular races that were thought to be scientifically classifiable through the detailed study of the human body. With growing criticism as to the viability of classifying the races through skeletal measurements, the practice of building collections of human remains seemed to have an uncertain future.

RACE, AGE, AND HUMAN ORIGINS RESEARCH IN THE AMERICAN ASSOCIATION OF PHYSICAL ANTHROPOLOGY

Despite the disturbing context of scientific racism in Europe during the 1930s and 1940s, of which the anthropological community in the United States was well aware, human remains continued to be used in comparative studies of race. Many scholars, in fact, did not share the misgivings that Boas and Cobb had regarding the project of developing strict and unmoving definitions of race. An example is the early history of the American Association of Physical Anthropologists (AAPA)—a history dominated by anatomists. At the first meeting of the association, anatomists authored nineteen of the twenty-nine papers.[64] The preponderance of anatomists reflected a continued, if at times uncomfortable, marriage between the fields of anthropology and anatomy. The lingering centrality of questions surrounding racial classification dominated the early history of the association, although it gradually gave way to questions regarding human history, as Cobb, Boas, and others attacked the notion that races could be strictly categorized.

When the AAPA met for a third time in 1932, the meeting was held at the Smithsonian. While a growing number of scholars trained in anthropology began to attend the meetings, the organization was still engrossed in the question of race. Papers presented at the meeting included "The Nose of the American Negro," "The Relations of the Sciatic Nerve to the Piriformis Muscle in American Whites and Negros," and "Dermatoglyphics in Shoshoni Arapaho Indians."[65] Again in 1935, with the organization growing in size, papers on similar subjects—including a study of Blackfoot craniology, a general study on the anatomy of "the American Negro," and a paper titled "The Plasticity of the Japanese Physical Type"—were presented alongside papers on human evolution. Papers comparing the anatomy of humans to that of primates were presented alongside a paper titled "The Roles of Undeviating Evolution and Transformation in the Origin of Man."[66] While the interest in human prehistory and evolutionary approaches at this conference anticipates future developments, the field was still, at this time, dominated by studies of comparative race. Many of these studies, those presented by both anatomists and anthropologists, were based heavily on research in bone rooms of museums across the country.

During the 1930s, a large number of scholars became increasingly interested in human growth and aging. The desire to understand aging came from a number of basic problems in both anthropology and archaeology. Knowing how human bodies age, specifically the aging process experienced by human bones and teeth, was thought to be useful for informing studies on diet and nutrition, as well as for providing archaeologists and anthropologists with data to calculate the age of fossils or remains discovered on archaeological sites. Similarly, archaeologists who encountered human remains were also eager for clues regarding the ages at the time of death of the skeletons they discovered. In 1936, the AAPA heard papers including "Changes in the Dimensions and Form of the Face with Age" and "Developmental Changes in Facial Features."[67]

By the mid- to late 1930s, human origins research had gradually become ensconced in physical anthropology research in the United States. In 1937, Robert Broom of the Transvaal Museum of South Africa traveled to the annual meeting of the AAPA to deliver a paper on new research related to the discovery of a distant human ancestor, *Australopithecus afarensis*. Broom's discovery made him a respected figure in the international anthropological community, and his presence in the United States was part of an ongoing, gradual shift away from those who focused more strictly on race toward those who were making impressive discoveries in places like Africa and Asia. For scientists in the United States, this shift seemed punctuated by the discovery of each new, impressive fossil emerging from places abroad.

Meetings in 1937 also witnessed scholars presenting papers that were more critical of comparative racial anatomy and the study of fragmentary human remains than in years past. Papers presenting racial classifications or the analysis of racial types continued to be presented; however, it was in organizations such as the AAPA where scholars from both museums and universities who were interested in comparative racial anatomy, human growth and diet patterns, and human origins and prehistory temporarily struck a balance.[68] This balance eventually tilted in favor of studies of human ancestry, but, for the moment, the two lines of study struck an uneasy accord. Bone rooms in major natural history museums and smaller medical museums around the country were unusually active, hosting scholars who were asking a wide range of large questions about humanity. Despite the increased attention paid to skeletal remains to answer a variety of questions, the limitations of research on human remains remained unclear.

THE LABORATORY OF ANATOMY AND PHYSICAL ANTHROPOLOGY AT HOWARD UNIVERSITY

W. Montague Cobb, in the introduction to his important monograph chronicling his career up to 1936, began by explaining to his audience, "These pages are a record of an attempt to keep the faith."[69] While Cobb enjoyed the start of a successful career in medicine, he witnessed many of the numerous challenges facing African Americans who attempted to enter into medicine and anthropology during this era. Lingering Jim Crow policies and racist attitudes led to the denial of job applications and fellowships and to discouragement from attending academic or professional meetings in many cities. Yet at Howard University, one of the nation's oldest historically black colleges, scholars like Cobb enjoyed the freedom to build rapidly professionalizing departments, shaping curricula and organizing new research programs. In 1932, Cobb was assigned the task of developing a program for the teaching of gross anatomy and physical anthropology at Howard's medical school, "the assumption being that this work would enable the Department as a whole to conform to the highest standard of medical education."[70] Cobb attempted to model the department after the established program at Case Western Reserve University, where he had spent time as a fellow working with its collections.[71] Cobb's work and career, too, were increasingly supported by a growing network in Washington that shared intellectual concerns in anatomy, race, and physical anthropology.

In developing the anatomical collections for Howard, Cobb built on existing collections. Older collections had been brought together by Daniel Smith Lamb, who acquired a large number of teaching models, mammal skeletons, preserved dissections, and bones that had been subjected to gunshot trauma. Lamb worked as a faculty member at Howard from 1873 to 1923, and when Cobb was first offered a chance to survey Lamb's collections, he encountered a striking sensation: "As I enthusiastically noted the scope of his interest, the meticulous attention to detail and the vision of future needs which specimen after specimen in surviving dust-covered boxes revealed, his purposes seemed so plainly evident that I felt I was reading his original thoughts."[72] Cobb's notion that he was reading the thoughts of the deceased scholar through an examination of the collection of human remains is telling and points to the lingering notion of these collections as both intellectually and physically constructed by determined individuals working at institutions around the country. By exploring the nature of collections, Cobb seemingly contends, the ideas and motivations behind the drive to collect, research, and display human remains could be better understood.

The collection that Cobb inherited was not without problems. In the years between Lamb's retirement and death and Cobb's arrival at Howard, some of the collections had either deteriorated or been lost during the move to a new facility in 1928.[73] Nevertheless, Cobb began his program for teaching and research at Howard with an existing collection of human and animal remains. Within a few years of his arrival at Howard, Cobb built on his knowledge of researching and displaying bodies, not only gathering literature from medical departments from around the globe but also examining numerous museums as references in crafting a plan for displaying human remains at Howard.[74] Ideas, in other words, were being disseminated through museum exhibitions as well as professional articles and monographs. With a trained eye, one might even sense the ideas being conveyed by these collections just by looking at them.

Within only a few years, Cobb had worked with other faculty to build an anatomy department including a dissecting room, an embalming room, and a morgue.⁷⁵ Exhibit cases spread throughout the building displayed a mixture of models, charts, and bones from human cadavers. Actual remains included a display of long bones and a pair of cases containing a wide variety of human skulls.⁷⁶ The exhibition was certainly more modest than were heavily funded displays like *The Story of Man through the Ages* in San Diego (1915) or those to come in *The Hall of Races of Mankind* in Chicago (1933), and it was intended for a different audience—medical students, as opposed to the broader public. Furthermore, the modest exhibit at Howard worked to break down existing schemes of racial classification more explicitly than did the other two examples, which largely underscored and tacitly promoted ideas of racial classification. All three exhibits, despite obvious difference in aims, utilized evidence drawn from human remains and anthropometry to build their underlying arguments.

The exhibits at the Laboratory of Anatomy and Physical Anthropology at Howard were not intended for broad public viewing. Rather, they were intended to teach medical students, who spent much of their time in the building. Cobb argued that exhibitions of this kind were a necessary and desirable component for teaching medical students: "The advantage of association of a museum with an anatomical laboratory, today need no argument."⁷⁷ Cobb noted that displays of this kind should include exhibits on human structure, growth and development, variation, prehistory, and phylogeny.⁷⁸ Cobb further articulated the role of the museum within the department of anatomy by explaining the process through which remains are acquired: "If material is carefully and fully utilized, a museum will inevitably result from the work of an anatomical laboratory."⁷⁹ Cadavers, in other words, would have continued use as objects for teaching in museum displays following a careful dissection. After surgical techniques were learned from the remains of the recently deceased, bodies could be appropriated as tools for understanding another series of questions about race, aging, and pathology.

While Cobb did emphasize teaching medical students about human difference in terms of race and gender, strict forms of racial classification were clearly deemphasized in the small exhibitions. In both Cobb's major text and his actual displays, the subject of racial difference was assigned roughly the same amount of space as sexual difference was, a fact that earlier intellectuals crafting medical museums would have found striking. Despite the deemphasis on racial classification, Cobb did note that one of the strengths of Howard's collection of human remains was its number of African American specimens. Instead of addressing the nature of Jim Crow medicine during the time period in which he was writing, Cobb delicately described the number of African American remains as a unique and positive feature of the collection, capable of drawing other researchers to Howard in order to study the remains.[80]

While Cobb's text generally deemphasizes racial classification, which he terms "racial anatomy," the subject is not wholly ignored. In describing the overall research program of the department, Cobb articulates some of the existing problems in the field of racial anatomy: "The study of racial anatomy has proceeded through the last century slowly but steadily like a stalwart in a storm. Always beset with influences which made for political bias, scientific method in this field has been hampered especially by the headline hunters of various groups interested in self-perpetuation, which have repeatedly snatched from students of human variation, unrefined data and immature conclusions for incorporation into their own ideology."[81] Jumping from the subject of the study of racial variation within Howard's research program, Cobb continues by addressing the lingering effects of racism and the "Negro Slave Trade" within the study of racial difference: "In respect to the American Negro, it was peculiarly unfortunate that the commercial possibilities of the slave trade were becoming manifest at the very time when physical anthropology was emerging as a separate discipline in Europe, during the latter part of the seventeenth and eighteenth centuries. At this time European knowledge of West African cultures was practically negligible."[82] Cobb continues by further linking the commercial slave trade to the rise of scientific racism:

> The three factors of commercial interest, ignorance and pride of conquest thus combined to create in the mind of the European civilization of the day an impression of biological inferiority as regarded the black man. There is little occasion for surprise that early physical anthropologists seemed to accept the concept of the stratification of human races, with the white race at the top, as biologically sound. Nor is it remarkable that there should have been instances where able men adduced anatomical evidence in support of this view, either because of sincere conviction, or, unconsciously, to furnish justification for a trade which currently represented powerful economic interests.[83]

Cobb's interest in downplaying anatomical racial difference was thus clear. He viewed himself as reacting to a historical context of scientific racism linked directly to the slave trade. Cobb continues his argument by pointing to several recent studies that worked to break down anatomical ideas of racial difference. Further, Cobb was unafraid to address the link between racial classification and the creation of a social stratification system designed to keep people of color from achieving the social, political, or economic status of whites. Specifically, he viewed these social patterns as continuing in some form from the slave trade to the present day, and the misguided and racist science of racial classification had only worked to uphold those existing social structures.[84]

CONCERNS ABOUT EUGENICS

Meanwhile, Hrdlička, who remained the preeminent physical anthropologist in the United States, became increasingly concerned with the direction that eugenics was taking by the early 1940s.[85] Though he supported the fundamental idea of eugenics, he struggled with the manner in which it was implemented, often critiquing official documents of the American Eugenics Society, an organization to which he had maintained a loose affiliation. By 1940, he wrote in a letter directly to the secretary of the American Eugenics Society, "The whole field of Eugenics is not at present in a very good state." He elaborated, "The fault lies in the fact that there have been advanced, as dogmas, various opinions and claims, before they were fully elucidated and sustained by science." Concluding his thoughts, he wrote, "The subject has become the prey of popular writers, and also some scientific propagandists rather than researchers. It needs a lot of young blood of the best kind so that it may be reestablished as a thoroughly high-class scientific procedure."[86]

Hrdlička's official departure from the eugenics movement came much later and with a much softer tone than many other anthropologists of his generation. Alfred Kroeber, for his part, had labeled eugenics "a joke" during a public lecture many years earlier, in 1914.[87] Unlike Hrdlička and the eugenicists, cultural anthropologists of the early twentieth century, many of whom ascribed to the ideas of Franz Boas, rejected the notion that heredity could determine innate ability; the goal of the anthropologist, they argued, was to document *culture*. The importance of these two major veins of anthropology cannot be understated; distinct groups of cultural and physical anthropologists had begun to emerge, and it was on these poles that the field ultimately was to be difficultly defined. At this time, however, the two groups worked alongside each other, albeit diverging intellectually as the years passed. By the late 1930s and early 1940s, many of the most fervent supporters of eugenics within the anthropological community were forced to face the realities of changing scientific and social theories against the dark backdrop of scientific racism emerging in Europe.

Despite the departure of Hrdlička, a leader in the field of physical anthropology, from the American Eugenics Society, the organization felt the continued presence of both physical anthropology and museums for decades. Harry Lionel Shapiro, who held leadership positions within the *American Journal of Physical Anthropology* and worked at the American Museum of Natural History and Columbia University, served as the AES president from 1956 to 1963. During the period of Shapiro's leadership of the AES, the African American physical anthropologist and civil rights activist W. Montague Cobb even spent some time on the organization's board of directors. This said, Cobb did not appear to be active within the activities of the society, and his motivations for joining the organization are unclear.[88] Ultimately, lingering associations with eugenics further harmed the perception of the project to establish human remains collections and their ethical treatment by scientists.

. . .

Between 1920 and the end of the Second World War, several factors led to crumbling scientific certainty that had once firmly buttressed racial classification schemes. Bone rooms helped define lines that supposedly divided races, but further studies and changes in the scientific community were blurring what at one time seemed so clear.[89] Well before the 1940s, most prominent physical anthropologists expressed concern with the eugenics movement—even as its significance in determining policy and social attitudes continued to expand. Racial classification was more and more understood to espouse inherently racist implications. The claim that human achievement could be tied directly to the human form was under steady fire. The study of prehistory and human origins, however, provided safer territory for many scholars, as these studies often appeared more removed from the obvious contemporary implications connected to race. Fueling these debates were arguments drawn directly from studies on human remains collections themselves. In the United States, anthropologists like Cobb and Boas utilized new discoveries to dispute the continued claims of racial difference made by scholars like Hrdlička and Dixon. Many scholars collecting remains abroad turned their attention to questions of human evolution, subsequently influencing the scientific discourse under way in the United States. Despite some shared agreement, these intellectuals disputed details surrounding measurements, methodology, and remains. In spite of a declining emphasis on the study of race, many viewed questions surrounding the subject as unresolved, thus requiring still more study of new remains. Infusing nearly all of these intellectual debates was an increasingly competitive drive to fill the bone rooms of museums in the United States with unique and valuable skeletal material from around the globe. Harvard and the Smithsonian became clear leaders in the field, but museums like the Field Museum, the American Museum of Natural History, and the University of California Museum of Anthropology followed closely behind, each reflecting particular regional strengths based on the interests of curators.

Scholars of prehistory and human origins increasingly utilized human remains collections during the 1940s—many times studying the exact collections used in earlier decades to investigate racial classification. World War II, like World War I, had a major impact on the collecting of human remains, but in a strikingly different manner—one that was in many ways intellectual rather than opportunistic. Scholarship in physical anthropology in some ways mirrored the course of the field of cultural anthropology. Several key cultural anthropologists, led by Franz Boas, were shifting the central focus of the discipline away from the study of race and toward the study of the concept of culture. Albeit in a slower transition, the study of physical anthropology, based heavily on research of human remains collections in museums, was shifting from studies centered on race to ideas revolving around human history.

Boas, reflecting on his recent campaign to shift the gravity of the discipline of anthropology away from the study of race, pointed to the growing global implications of studies in racial classification. In a letter appealing for grant funding, he wrote,

> For a number of years I have been engaged in investigations relating to racial characteristics, particularly for the purpose of showing the lack of any scientific basis for the theories which are at present dominant in Germany. I dare say that largely owing to my investigations the general position of American scientists, who ten years ago were dominated by racial enthusiasts, . . . has completely changed and that it is generally recognized that social factors are infinitely more important than any so-called racial hereditary characteristics. It will take time to have the general public understand this, but we are doing out very best to make our stand well known.[90]

While Boas and other scholars who followed him began to reflect on the intellectual shifts taking place during the past decade, they were faced with the history connected to collecting human remains, which left behind an indelible mark on museums across the United States. Many museums found their collections growing despite criticism. As bone rooms expanded, the questions that skeletons were presumed to answer began shifting and continued to evolve over the ensuing decades.

CHAPTER 6

SKELETONS and HUMAN PREHISTORY

Forty years after the wildly successful World Columbian Exposition, Chicago again hosted a world's fair. The 1933 celebration came to be known as the Century of Progress International Exposition. Race and history again loomed large as subjects in fair exhibits. This time, however, such exhibitions received more space than at the 1893 fair in the same city. Emerging in the exhibitions was an increasingly balanced emphasis on the comparative study of race alongside a growing project to utilize bodies as tools for conveying ideas about human evolution and prehistory. Human remains collections that had been previously languishing in bone rooms progressively became more central to museums' exhibition strategies, teaching visitors about both race and human evolution—themes often connected but sometimes exhibited separately. Anthropologists in the era mostly agreed with Fay Cooper-Cole when he described the fair's return to Chicago as "an unusual opportunity to present Anthropology to the general public."[1] How best to make such public presentations, however, became a matter of considerable debate. Voices from the academic community, like Cooper-Cole, sounded against the curators of museums and the dwindling number of entrepreneurs who hoped that displaying their own collections at the fair would bring in a quick buck. Debates about exhibits reflected an ongoing shift in the study and display of human remains in museums in the United States that was taking place in the late 1920s and early 1930s. Museums in this era began to move away from previously emphasized exhibits on scientific classification of the races toward the study of prehistoric humans and human evolution. Before this time, ideas presented to the public about racial classification had been largely noncontroversial. With increased attention to competing ideas about human origins in the mid-1920s—particularly with the Scopes Trial's publicity—exhibitions displaying the suddenly controversial ideas related to human history grew in size, complexity, and scope, while at the same time garnering significant public interest.[2] Rather than simple displays of skulls and bones under glass cases, these exhibits started to tell stories. The stories told in the exhibits—through both scientific and artistic interpretation of bones and fossils—was sometimes embellished and even falsified in order to construct coherent narratives for eager museum audiences.

Cooper-Cole, a University of Chicago professor, led the drive to organize official anthropological displays for the fair in the early 1930s. Unfortunately for Cooper-Cole, his displays were overshadowed by two large, new, permanent exhibitions at the Field Museum of Natural History in Chicago. These competing exhibitions were organized by an energetic and well-connected curator of physical anthropology, Henry Field. Field had previously achieved some success displaying human remains. His exhibitions were heavily influenced by the earlier displays organized by Hrdlička in San Diego in what became the San Diego Museum of Man. Field hoped to expand on what was presented in San Diego while adding in his own unique flair for the dramatic and improving the artistic renderings of prehistoric scenes in museum displays. The Field Museum exhibits, bringing together contemporary ideas about race and history, represent a snapshot of ideas drawn from bone rooms just before the decline of racial classification studies and before physical anthropology began to focus more heavily on history and ancestry.

Immediately before the decline of scientific racism, existing human remains collections emerged as an important tool for studying the deep human past. Scholars interested in studying human remains began to change their language from one centered on *race* to discourses surrounding *population, migration,* and *evolution*. Many began to argue that attempts to understand the complex nature of human diversity should not be used to create a ranking system for various human races; instead, ancient skeletons were most useful in solving the riddles of the human past. This line of thinking did not convince everyone, and certain anthropologists continued to promote traditional forms of racial science. Modern molecular genetics was, before the Second World War, only barely in its infancy, and scholars were unsure about the future of understanding lineage through new technologies.

The supposed fall of scientific racism in the United States proved to be partial and incomplete. In describing the climb of the Anglo-American male to the apex of human civilization, many sources in the era positioned the development of the European-American ideal in relation to both human evolution and historical progress.[3] Further, the public remained largely convinced that physical differences as reflected in bones might point to inherent abilities or disabilities directly linked to race, not to mention sex. Museum exhibits, such as those opening at the Century of Progress in 1933, likely influenced these and other popular ideas.

When the physical anthropologist T. Dale Stewart reflected on his own field in the mid-1970s, he made an important observation about the history of human remains collections. In an article published in *Anthropological Quarterly*, Stewart argued that physical anthropology—unlike ethnography, archaeology, and linguistics—remained museum centered several decades longer than did these other fields, which had mostly witnessed a shift in the major production of new ideas away from museums to university campuses years before the 1940s.[4] Reconsidering the history of human skeletal collections, physical anthropology, and the study of human origins in this era provides new insights about the twilight of the "museum period" in the United States, as the nation continued the work of establishing and building permanent museums for both the public and the benefit of science.[5]

MAGDALENIAN GIRL

Although popular audiences had seen numerous representations of the human form as well as actual human remains in museums and fairs before the mid- to late 1920s, the vast majority of these displays focused on racial typing. In the late nineteenth century, as naturally mummified remains discovered in the American Southwest became popular for exhibitions, displays of human remains, even the most ancient, were heavily racialized. Earlier, I examined how the emerging concept of "racial history" worked to blend these two uses for human remains. Despite the existence of mummies in the American Southwest, as well as the emergence of new pre-Columbian remains from around the Americas, scholars and the public at large continued to associate the ancient with Europe, Egypt, and the Fertile Crescent region of modern-day Iraq and Iran. Because these narratives were more closely associated with the rise of modern Western civilization, the emphasis placed on the display of prehistoric bodies in the United States was less racialized. Casts of these finds existed in museums in the United States; but originals rarely traveled outside their countries of origin, and exhibits about human evolution remained infrequent. New discoveries, coupled with a dramatic increase in debates about human evolution in the 1920s, continued to bring both popular and scholarly attention toward research and display of ideas related to human evolution.⁶

In the 1920s, the debut of a prehistoric skeleton discovered in France captured the imagination of the Chicago media. The display was organized shortly after Henry Field at the Field Museum of Natural History acquired a nearly complete skeleton in southwestern France, dubbed "Magdalenian Girl" after the so-called Magdalenian portion of the Upper Paleolithic period, about fifteen thousand years ago. Magdalenian Girl was smuggled from the Cap Blanc region in France during World War I, reportedly in a coffin disguised as containing a fallen American soldier. The remains were transported to the American Museum of Natural History in New York before being sent to the Field Museum. Records indicate that the Field Museum agreed to purchase the remains for the price of $12,000, but an official transaction never seems to have been finalized. Field subsequently acquired a projectile point purportedly discovered near Magdalenian Girl's remains. Before the skeleton went on display, he encouraged the local media to speculate on the nature of the death of "Miss Cro-Magnon." Perhaps the projectile point discovered near her body was used to kill her, the press vividly reported, and maybe this murder had occurred at the hands of a jealous lover. Archaeological evidence confirming these claims was limited at best. Nevertheless, when the remains went on display in the mid-1920s, the Field Museum smashed a single-day attendance record as twenty-two thousand spectators came to view her skeleton.⁷

New discoveries related to prehistory fueled growing desires to see prehistory represented on display. Further fueling this urge, the press elaborated on popular imagination surrounding ancient remains—fantasies often including sex, love, and untimely death. Eventually, new displays created between the late 1920s and the 1930s pushed prehistory as a concept to share the stage with displays on the subject of racial classification. The showmanship surrounding the display of Magdalenian Girl was reminiscent of the tradition of P. T. Barnum, and other exhibitors similarly took advantage of popular interest in ancient human bones coupled with a good story.[8] Over the course of the ensuing decade, however, those displays related to race and prehistory grew larger and more complex. The growing number of displays on human evolution continually embraced fantasy and imagination as a method for teaching public audiences that toured exhibitions. But the new exhibits also served another purpose: they worked to solidify the museum's place within scholarly discourse about race, human history, and human remains. Strikingly, this growing contribution to the discourse began in earnest just as many scientists were abandoning museums, with many of the best minds leaving to teach and study on university campuses.

A CENTURY OF PROGRESS

In the early stages of planning anthropological exhibitions for the 1933 world's fair, competing interests became clear. Fay Cooper-Cole, who helped organize the anthropology department at the University of Chicago, hoped to enlist Aleš Hrdlička from the Smithsonian Institution to create displays under the auspices of the official fair exhibitions. Cooper-Cole wrote to him simply, "I am sure we are both agreed that Physical Anthropology should have an excellent exhibit and I want you to have charge of it."[9] Problems became clear in 1931, however, when Cooper-Cole learned that the Field Museum of Natural History also wanted to open new anthropological exhibits coinciding with the fair. Specifically, Cooper-Cole caught wind that the Field Museum had already begun the process of planning new exhibits on race. Cooper-Cole was thus eager to start planning for the displays at the fair, which he believed to already be falling behind the other exhibits competing for attention. Compounding his problems, he thought, were issues of clarity with regard to how the proposed fair exhibits would address race and prehistory; "Physical Anthropology should plan its exhibit," he argued, "so as to cause a minimum of overlapping with pre-history."[10] Hrdlička had earlier proposed exhibitions following humans from the early embryonic stage of life to old age, as well as an exhibit that would "follow man from his early types to the modern races."[11] These proposals proved intellectually problematic. The plans for the fair already included proposed displays on pathology under the auspices of a medical exhibit. Hrdlička eventually turned down Cooper-Cole's offer to organize displays related to physical anthropology for the fair. The problems that Cooper-Cole encountered in constructing exhibitions for the 1933 fair were in many ways representative of problems encountered by the field of physical anthropology as a whole and by others studying collections of human remains in the same era. The lines between the study of evolution, medical pathology, and race had solidified in terms of disciplinary professionalization, and yet many of their ideas overlapped in a confounding manner.

Cooper-Cole was also working to make striking exhibitions from scratch in a city that now boasted a large natural history museum. By 1927, the Field Museum of Natural History already oversaw a storehouse of about three thousand sets of human remains. The vast majority of the skeletons, however, had never before been seen publicly.[12] While Cooper-Cole struggled to articulate an exact vision for original fair exhibits, curators at the Field Museum had for several years envisioned two new permanent exhibits on race and prehistory. Henry Field, who became curator at the Field Museum following his doctoral work at Oxford University, was trained in the brief moment when race and prehistory held an almost equal standing in anthropology.[13] Henry Field worked and traveled tirelessly, promoting both his ideas and his adventures to other scholars around the world. Field eventually left museum anthropology to take a position in the federal government, having never secured a place as a major anthropological thinker in a generation that experienced radical shifts in the field.[14] Nevertheless, during a brief period in the 1920s and 1930s, Field pushed displays of human remains or displays based on the study of remains in new directions. Upon his arrival at the museum in Chicago, Field began working with other scholars both in the United States and abroad, collecting avidly with the desire to create specific displays on prehistory.[15] The 1933 fair simply provided an opportunity to display existing ideas. Field worked to develop the two major exhibits based largely on physical anthropology, splitting race and prehistory as themes, with both halls opening in time for the 1933 Century of Progress fair.

The process of organizing *The Hall of Races of Mankind* was more complex than was the development of *The Hall of Prehistoric Man*, but progress in creating the two exhibitions was connected in certain ways. Field decided that the best approach to building the exhibition about race was to begin with an extensive review of the available literature surrounding racial classification. In particular, the issue of exactly how *many* races to depict in the exhibit was an ongoing point of debate. The head of the Department of Anthropology at the Field Museum, Berthold Laufer, suggested that Field travel to San Diego to examine the exhibitions organized earlier by Aleš Hrdlička. Field ultimately traveled to a number of different museums in California, but the museum founded on the work of Hrdlička and Hewitt was the most important in shaping his thinking about the forthcoming exhibitions in Chicago. Arriving in San Diego in 1930, Field encountered a relatively young museum, at the time called simply the San Diego Museum, born out of the Panama-California International Exposition's *Science of Man* building. *The Story of Man through the Ages* had started to show some age but was still considered the only major exhibition of its kind in the entire United States. The museum had even changed locations since its founding, moving most of the original displays to a new building. Reconstructions of prehistoric people by Aimé Rutot and busts depicting races of humankind modeled by Frank Micka, in particular, remained on exhibit in San Diego. The museum also prominently featured human remains collection in exhibits, drawing audiences to galleries as the museum grew into an independent institution.

Field was generally impressed with existing displays in San Diego and took extensive notes about the galleries—even hiring a watercolor artist to duplicate the color graphs on exhibit during his stay in California.[16] Describing the displays as now somewhat outdated, he observed that the museum in San Diego remained among the most complete exhibitions on either race or prehistory in existence. The content of the displays had shifted only slightly since their organization decades earlier, but the order of the galleries had been overhauled since the museum's opening. Field looked to push further the ideas contained in these halls in new exhibitions in Chicago. In recounting his visit to the exhibitions in San Diego, he recounted, "I obtained many new ideas and suggestions dealing with the exhibition of material relating to the races of the world, not only from an educational standpoint, but also from the point of view of the average intelligent museum visitor." But Field was not just impressed with the San Diego Museum's presentation on race, adding, "The series of prehistoric reconstructions . . . is fantastic in the extreme."[17]

Although Field believed that the presentation of ideas about both race and prehistory was strong throughout the San Diego exhibit, he was also critical. Not only were many of the charts and graphs dated, but he also found the artistic rendering of different races to be "poor" and the "facial expression" in the busts "almost entirely lacking in every case." Field observed that the human remains displayed in the San Diego exhibitions were in certain ways strong, but he criticized the displays for not including any associated flora and fauna. In his view, skeletons alone were not enough to fully convey ideas about human evolution and prehistory. Instead, remains or reconstructions of early humans needed to be artistically depicted in some context. For museum curators and exhibit planners in Chicago, this meant creating new dioramas. Field described the reconstructions in San Diego: "The method of showing casts of the human remains accompanied by restorations is excellent but the latter here are so poor that the resultant impressions are totally erroneous." Field hoped that Chicago might learn from the presentation of complex ideas about race and prehistory in the San Diego displays, while expanding on the ideas, updating them, and also working to improve the artistic rendering of complex ideas about both subjects.[18]

After Field had studied the literature and corresponded with other anthropologists, he came to the idea that exactly 155 racial types existed. He soon changed his mind, determining that 164 racial types existed on earth.[19] In crafting the exhibitions for the 1915 San Diego fair, Hrdlička had emphasized creating a single, unbroken line of busts representing the racial diversity of humankind. Were the uninterrupted line of statues somehow broken apart, Hrdlička maintained, "that would destroy all original sequence and change the beauty of the exhibits."[20] *The Hall of Races of Mankind* clearly modeled this approach, employing an unbroken line of full-sized sculptures and smaller busts, surrounded by clean, stark walls, thus forcing the visitor's gaze toward each statue. Personally, however, Field believed that the artist he was preparing to hire to model the races of humankind for the exhibits in Chicago could outdo the artist commissioned for San Diego, Frank Micka. Further, he believed that by creating more extensive dioramas including reconstructions of early technology and the surrounding environment, the presentation of human prehistory would be more complete and better understood by visitors.

Henry Field had been a great admirer of Aleš Hrdlička and the Smithsonian for some time. In 1926, upon Field's graduation from Oxford and before his arrival in Chicago, he briefly visited Washington, DC, and viewed the Smithsonian's collections firsthand. Field paid close attention to the manner in which the national museum stored and organized the remains, noting, "Your methods of arrangement and indexes have been my admiration since leaving you and I shall certainly try to model my department along your lines."[21] Over the next several years, Hrdlička advised Field on bone classification and storage, as well as on the subjects of research and display.[22] When Field was finally ready to propose a layout for new exhibition spaces to his supervisors at the museum, he had clearly benefited from Hrdlička's ideas regarding racial classification, management of human remains collections, and exhibitions.

To turn Field's vision for an expansive exhibit on racial classification into a reality, the museum hired a talented artist named Malvina Hoffman to travel the world and create lifelike busts and full-sized bronzes depicting agreed-on racial types. Hoffman was as gifted as a negotiator as she was a talented sculptor. After conferring with Field Museum officials, Hoffman and Henry Field agreed that the exhibition would consist of twenty full-length bronzes, twenty-seven busts, and one hundred additional head figures. Hoffman was awarded an unprecedented contract of $109,000 for her work and an additional $125,000 for her travel expenses.[23]

The Hall of Races of Mankind was hugely successful in both attendance numbers and overall popular reception. Visitors to the museum eagerly purchased enough exhibit catalogues to require printing several editions. The manner in which racial lines and categorizations were presented in the exhibition was strict, making the ideas generally accessible to broader audiences. In the preface to the exhibition catalogue, Berthold Laufer pointed to the exhibition's emphasis on rigid racial types: "As a biological type our

Negroes belong to the African or the black race and will always remain within this division; even intermarriage with whites will not modify their racial characteristics to any marked degree."[24] The clarity of the exhibition, which presented solid and largely unchanging racial classifications represented by singular or small groups of elegantly constructed statues, encouraged an understanding of race that fit neatly within American popular consciousness. Further, the exhibition blended the science of studying the human form with the talents of a gifted artist, accomplishing an aesthetic rarely achieved by natural history museums in the United States before this point. Sir Arthur Keith, who served as Henry Field's primary mentor, also penned the introduction to the exhibit catalogue, explaining his view on the central problem that museums had encountered in presenting the subject of race: "How can such a vast assortment of diverse individuals be given a true and effective representation in a museum? According to established precedent, human skulls, skeletons, photographs, charts, casts, and models brought home from all lands fill the exhibition cases of such a hall in museums. And such collections no doubt prove of great value to professional students of anthropology, but exhibits of this nature are likely to repel rather than to attract visitors to the study of mankind."[25]

Although Keith does not mention the displays brought together for the Panama-California Exposition in 1915, he might just as well have been comparing the two exhibitions directly.[26] Keith's statements are noteworthy in that they follow a pattern in arguing that the public was repelled by human remains when in fact audiences had repeatedly proven otherwise. Nevertheless, Keith's notion that some members of the public were *repelled* by actual displays of human remains yet *drawn* to artistic representations of conclusions drawn from the study of those remains became the dominant strain of thinking over ensuing decades. The exhibits brought together by Hrdlička and Hewett for the 1915 fair—which served as reference and inspiration for Field's Chicago exhibits—were largely the type that Keith describes: skulls, skeletons, charts, and photographs, with heavy glass cases. The layperson, Keith argued, found human skeletons repelling, while, in contrast, the white walls and strong, artistic bronze statues made the study of race attractive. This argument was, in fact, far too simplistic. Visitors had been attracted to the display of human remains for generations. While the public may have shown some revulsion to the display of bodies, they filled museum halls to see them since the middle of the nineteenth century. Keith's remark make it appear that he was unaware that the single-day attendance record at the Field Museum was set by visitors who wished to view a prehistoric human skeleton. Despite evidence suggesting otherwise, Keith's contention that the direct display of human remains would repel, rather than attract, visitors was a concept, curiously, that continued to be prevalent in museum thinking for years. Later resistance to the display of human remains combined with the ethical challenges to exhibition of ancestral remains essentially worked together to remove most human remains from display altogether.[27]

The Hall of Races of Mankind was arranged by both racial type and geography. Humankind was at once a "well-defined uniform species" and also a species rationally divided into groups. The evidence for racial categories, explained in the catalogue accompanying the exhibition, was based in the study of "the physical characters of the living person, and the anatomy of the skeleton."[28] Though detailed measurements of living humans and the close study of human remains provided the evidence on which the exhibition was based, it was clear that this evidence was not being presented in this particular gallery. Here, visitors were free from being asked to engage with the skeletons they might find abhorrent, instead viewing cleaner bronzes meant to represent the measurements of the living and the dead.

In Keith's introductory remarks to *The Hall of Races of Mankind*, he posed the question, if modern human types existed in ancient Egypt five thousand years ago, how could it be that humankind has changed at all over the course of history? Keith responded, "I would ask my critics to go back 50,000 years, and see what then? The answer is given in Hall C at the Field museum, which was devoted to the Stone Age of the Old World. The prehistoric human types exhibited there differed profoundly from their modern representatives."[29] Once visitors observed contemporary human differences through artistic renderings, they might wonder how human beings had developed in such a manner. It was assumed that visitors would naturally turn their attention to an adjacent hall that was grounded in the concepts of human evolution and prehistory. Although the two halls were distinct, they were meant to address the lingering questions that a curious visitor might have about humankind.[30]

The main attractions in *The Hall of Prehistoric Man*, ultimately, were the impressive and dramatic dioramas featuring hand-modeled human ancestors cast against delicately painted backdrops and set among reconstructed vegetation. Field cleverly utilized existing museum collections depicting Swiss lake dwellers, acquired at the conclusion of the World's Columbian Exposition, as well as authentic artifacts from the Lower Paleolithic from Africa and Asia, to complement *Homo erectus* models. Neanderthals, or *Homo neanderthalensis*, were depicted in what was believed to be an anatomically accurate, hunched-over stance. Models of human ancestors were not only displayed holding genuine artifacts but also featured lifelike poses and actual human hair.[31] A brief article in the *Science News-Letter* notes that while dioramas were the focus of the exhibit, the displays did include "reproductions of important specimens of prehistoric human remains as well as some original skeletal material, and fossil specimens of the animals of each period."[32]

Field maintained that he had visions for *The Hall of Prehistoric Man* dating back years. When the exhibit opened, he said, "The Hall of Prehistoric Man was all I had hoped it would be, as I had dreamt it since my sixteenth year. Here within the space of a half hour, walking past the eight dramatic and colorful dioramas, a visitor might read in true-to-life chapters the past quarter of a million years of Man's history."[33] In his even more hyperbolic writings, Field described fantasies for curating an exhibit on prehistory at the museum since his teenage years, yet professionally he recognized that his ideas were also shaped by more recent influences including other museum displays on race and human history.

Despite Field's personal tendency to inflate the importance of certain activities, both the popular and academic reaction to the new permanent exhibits were, in fact, overwhelmingly positive. While the exhibit on race captured the most popular and critical attention, the new displays on prehistory were significant in their own right. In particular, the continued display of the Cap Blanc skeleton, or Magdalenian Girl, was viewed as important for the museum; one article described the skeleton as "one of the most important archaeological treasures in this country." Detailing the creation of the prehistoric displays at the Field Museum, the article described the dioramas as "the finest restorations of prehistoric men ever made." Further, the exhibits were noted to be "the most complete, accurate and interesting picture that present knowledge permits of the lives, cultures and physical characters of prehistoric races."[34]

Compared to previous displays stemming from studies on human remains, the two Field Museum exhibitions opening in 1933 were more balanced in their dual concerns with prehistory and racial classification. Despite the physical—and consequently intellectual—separation of the displays into different halls, popular newspaper accounts conflated the ideas of race and prehistoric humans in reviews. Indeed, the museum hoped visitors would turn from one exhibition to the other *despite* the distinction of two separate exhibit halls. In describing the Field Museum's diorama displaying reconstructions of a Cro-Magnon scene, the *Chicago Daily Tribune* stated, "The third scene represents Cro-Magnon men of a race which invaded Europe from Asia about 30,000 years ago who are believed to be the first direct progenitors of modern races."[35] Unanswered questions about what these supposed races became were addressed only in the adjacent hall.

Although the Field Museum's exhibitions were largely a success, they were not without problems. *The Hall of Prehistoric Man* prominently featured a display on a purported fossil that turned out to be a forgery, Piltdown Man. Displays of Neanderthals were based on early discoveries of *Homo neanderthalensis* remains that happened to be arthritic. The Piltdown Man displays were removed in the early 1950s, and the exhibition was updated in 1972 and again in 1985 before finally being dismantled in 1988.[36]

The opening of *The Hall of Prehistoric Man* and *The Hall of Races of Mankind* represents a snapshot of the ideas surrounding the subjects of race and prehistory in the early 1930s. Henry Field, despite his lack of depth as a scholar, developed an ability to leverage connections in both the United States and Europe to create two wildly popular and influential exhibitions. In studying the displays created for the Field Museum, it might be noted that few actual human remains were on display at the institution; indeed, part of the appeal of *The Hall of Races of Mankind* was its clean aesthetic of bright white walls set against beautiful bronze sculptures. And while *The Hall of Prehistoric Man* did display both actual human skeletal material and numerous casts, the increasing dependence on dioramas and artistic reconstructions represented a break from the earlier displays of skulls and mummified remains.[37] What the displays lacked in macabre appeal they made up for with aesthetic improvements. The exhibitions that were brought together at the Field Museum in time for the 1933 world's fair are demonstrative of the fact that the emergence of the study of human prehistory largely arose from research on human remains. While scholars remained confident that human bodies could be classified into groups of different races, this theme began to share the spotlight with discoveries related to the study of human evolution.

RACIAL CLASSIFICATION, EXHIBITION, AND THE PUBLIC

Racial classification in the late 1920s and early 1930s did not always comprise complex charts, graphs, or meandering exhibitions highlighted by sophisticated artistic attempts to render humankind's many races. Scholars were, in fact, willing to break down ideas into more simplistic categories in what they viewed as an effort to enlighten the public. The manner in which these ideas about racial classification reached their apex in the displays of the 1930s, however, is apparent in writings and correspondence between scholars and members of the public.

When Herman J. Doepner, a man from St. Paul, Minnesota, wrote to Aleš Hrdlička asking for "a detailed modern classification of the races of mankind," he added that he desired a list of books on the subject and asked that the information provided be brief in order to "obviate much reading."[38] Hrdlička, who would typically reply to similar requests with lengthy, somewhat pedantic, bibliographic lists of writings on particular subjects, on this occasion wrote a rather straightforward reply to Doepner. Hrdlička lamented, "There is no satisfactory recent publication which would give the classification of races according to our latest knowledge." This said, he wrote, "But as a classification is rather simple, until we come to details, I will give it to you herewith": "We recognize today three main races or stems of mankind, which are: the White, the Yellow-brown, and the Black; with a secondary fourth group constituted by the Australo-Tasmanians. The Whites in turn are divisible in main into the Nordic, Alpine, Mediterranean, Semitic, and Hamitic types. The Yellow-browns embrace the Mongoloids, Malays, and aboriginal Americans. The Blacks compromise the Negritos and Negrillo; the Negro proper; the Bushmen and Hottentots; and the Melanesian blacks. Besides which there are the mixed Polynesians, and other smaller groups."[39]

Hrdlička's conceptualizations were important in framing the displays at the Field Museum concurrent with the Century of Progress Fair of 1933, though Henry Field credited his own work, as well as the work of other scholars in anthropology. Scholars working with the Field Museum largely embraced these strict racial classification schemes, as described in the passage on the exhibit quoted earlier, but they also simultaneously espoused rhetoric concerning the "unity of mankind." In Chicago, *The Hall of Races of Mankind* concluded with a massive bronze statue featuring men of three different races holding a towering globe. The modernist sculpture worked to classify and separate while also underlining the related nature of the human family.[40]

Notwithstanding Fay Cooper-Cole's initial lack of success in building exhibits, he was able to convince Harvard University's Peabody Museum to temporarily contribute a working anthropometry laboratory. Over the course of the fair, the laboratory acquired detailed measurements of around thirty-five hundred visitors.[41] Despite the overall decline of popularity of racial classification in the field of anthropology in the 1930s, the practice of anthropometric measurements constituted a significant aspect of both research *and* display of the period.

Just as museum displays began focusing on the unity of humankind, the world was starting to fracture. Earlier in 1933, Hrdlička wrote directly to Franklin Roosevelt on the subject of Japan. (Hrdlička had corresponded with Roosevelt earlier, when the latter served as the assistant secretary of the navy.[42] In 1933, when Roosevelt became president elect, Hrdlička again wrote to him about Japan.) Hrdlička considered himself in a special position to address the growing problems in the Pacific. He wrote, "I have endeavored, in particular, to learn as much as possible about the soul of the different peoples." Hrdlička's trips to the field had brought him to Japan, Russia, and China. He believed that the differences between these nations were not, at their core, due to essentialist racial differences. Instead, they arose from political circumstances. Nevertheless, the manner in which Hrdlička attempted to describe Japan's political behaviors is rife with racialist undertones: "There has been arising an ever more threatening obstacle, which is Japan. Not the Japanese people, who have enough good in them, but that something utterly egotistic, tricky and ruthless to the weaker, which is the governing clique of that country." Hrdlička continued, "This power since fifty years is working steadily toward to exclusion of all and particularly the white man from the Pacific, toward the domination, by hook or crook, of all it can reach, and towards a reign backed by brute force and all other means, moral or immoral, over all eastern Asia and the whole great Ocean."[43] The contrasts appear striking to the contemporary reader. Just as museum exhibitions balanced the tension between racial types and human unity, scholars were questioning the political behavior of nations in scientific rhetoric laced with racism. In Europe and in Asia, extreme political regimes based much of their rhetoric and power on exploiting various racial and cultural tensions. Exhibitions in the United States, meanwhile, based on the research and display of human remains, continued a transition from galleries focused largely on racial typing to exhibitions that split the stage between the study of race and of human history. The looming war forced scholars in the United States to face much more critically many of the tensions they attempted to balance in prewar exhibitions.

COLLECTING HUMAN ANCESTORS

In 1925, while looking through a box of seemingly random rocks and fossils in South Africa, the Australian-born anatomist Raymond Dart made a major discovery. He found buried in the box the fossilized fragments of a small skull discovered in a nearby quarry. Dart had been collecting fossils for several months, but the unusual find captured his—and soon worldwide—attention. These fossils turned out to be important; and yet Dart, an anthropologist and anatomist, was heavily criticized for his claims that the find represented an evolutionary link between contemporary humans and their distant ancestors. Dart's discovery consisted of the fossilized brain and cranium of a distant human ancestor that he named Taung Child.[44] The discovery of a human ancestor was not the first since Neanderthals were discovered in the mid-nineteenth century. An independently driven Dutch anatomist working with colonial support, Eugène Dubois, discovered a fossilized skull known as Java Man in Indonesia in 1891, for instance. While Dubois's discovery attracted comparatively little attention during his lifetime because Java Man did not fit into preconceived notions of the "missing link," Dart's discovery quickly captivated audiences in both Europe and the United States.[45] Hrdlička wrote to Dart one year later, in 1925, congratulating him on his find and expressing interest in publishing his account of the discovery, though Hrdlička assured Dart, "I presume that you will have no difficulty in its publication." Hrdlička moved quickly to reprint Dart's findings in the *American Journal of Physical Anthropology*. By the 1920s, Hrdlička had come to believe that further finds of significance would come from Africa: "I have no doubt but that now, since interest in finds of this nature has been so vivified, there will come to us many specimens of value from your continent."[46]

Africa, in the words of Dart, was "very young anthropologically." Though significant discoveries were being made on the continent, scholars had little in the way of resources. "We have great material," Dart wrote, "but we have little facility for securing it and consequently most of it is lost, and what is gathered is secured unscientifically." Dart invited American scholars to visit South Africa to contribute to the study of early humans.[47] Despite evidence to support his claims, Dart proved less adept at making the case that evolution of the human species had primarily occurred in Africa, and scholars in the United States and Europe remained skeptical. Other scholars working in the same period, including a Canadian physician named Davidson Black, made discoveries of different kinds of human ancestors that appeared to be closer to modern humans than did Dart's primitive Taung Child. Black worked to advance discoveries made in China around the same time, culminating with the discovery of a fossilized skull known as Peking Man, later understood to be *Homo erectus*.[48] Both Black and Dart eventually traveled to England to make the case for the importance of their discoveries, and Black was much more effective in promoting Asia as a site for human origins than was Dart in his promotion of Africa.[49] The European scientific community's focus on some scholars over others meant that even discoveries of clear significance to the story of human origins took decades to be recognized as such.

When Hrdlička and Dart began corresponding, scholars in the United States remained divided as to the most important location for studying human evolution. When working outside Europe or North America, scholars interested in researching or collecting human remains scoured the globe for fossils in Asia, the Middle East, and Africa. Discoveries of fossilized remains that were important to understanding human history occurred throughout the 1920s and 1930s. In Asia, the discoveries of Java Man and Peking Man promoted the region as a possible location of a "missing link." In South Africa, the first example of a much more primitive-looking adult australopithecine was unearthed in 1936 by the paleoanthropologist

Robert Broom. A decade earlier, in 1926, a British team, Louis and Mary Leakey, began their work in eastern Africa. The Leakey family's presence on the continent continued into the next century, training new scientists and continually announcing new discoveries that forcefully shifted the focus of international paleoanthropology to Africa.

Despite gradually revealing new discoveries, progress in studying human evolution remained frustratingly slow.[50] Before the 1950s and early 1960s, scholars in the United States and Europe could not agree on the continent that held the origin of human ancestors.[51] Fossils discovered in Africa or Asia often took years to be unearthed and subsequently described to scholars in the United States and Europe. Scholars most commonly communicated new finds through letters and publications in their home countries, and scholars in Europe, the United States, Asia, and Africa worked to cobble together pieces of the complex story of human evolution through limited evidence spread throughout museums and universities thousands of miles apart. Fossils went missing. Replicas of fossil specimens, or casts, were created, duplicating the traits of important finds, but scholars complained that these casts were often too poorly crafted to be accurate representations of the originals. Illustrations of original fossils, too, helped scholars in the United States understand the growing fossil evidence, but, like casts, the accuracy of illustrations was dependent on the skill of the artist attempting to represent the original fossil.[52] When the Academy of Natural Sciences of Philadelphia celebrated its 125th anniversary in 1937, the institution hosted a symposium on early humans. Reflecting the geographically scattered nature of the study of human prehistory, scholars from China, Java, South Africa, England, and Denmark were listed among the attendees. Materials promoting the symposium called for responses to the inherent problems in the study of early humans before World War II: "Broadly speaking, the objective of the Symposium and the attendant activities is to focus scientific attention on the advances being made in research on Early Man throughout the world and by correlating all pertinent information dealing with the broad but important subject, to enrich the scientific knowledge of results already obtained and to lay the foundation for a better correlated attack on the problem in the future."[53]

Along with the geographic diversity of the symposium, it was strikingly interdisciplinary for its era, featuring scholars from physical anthropology and archaeology, as well as geology and paleontology. Papers delivered at the symposium examined new discoveries related to the Folsom culture in North America, ideas about the Bering Strait land bridge, and new hominid discoveries in Africa and Asia.[54] The symposium served as a snapshot of current research surrounding human prehistory and human evolution but also provided a model for the direction of the field of physical anthropology, which was becoming increasingly interdisciplinary and taking a greater interest in the distant past as opposed to strictly racial classification. Just four years after the opening of *The Hall of Races of Mankind* and *The Hall of Prehistoric Man* at the Field Museum in 1933, the balance displayed between the study of race and of human history was rapidly shifting in the scholarly discourse.

One of the speakers at the symposium was a young scholar named Theodore McCown. The career of McCown, who went on to earn a doctorate from the University of California, Berkeley, in 1939, was somewhat representative of the struggles facing scholars interested in human evolution during this period. McCown's story is also rather unusual in several respects. As McCown studied for his doctorate, he began working with an archaeologist and physical anthropologist in the United Kingdom, Sir Arthur Keith. Working with Keith, McCown collected and began to study a series of fossils from Mount Carmel in present-day Israel. McCown battled increasing pressure to publish the results of his studies and was frustrated by an inability to conclusively date the fossils with the available methods.

Following an extended stay in what was then Palestine, McCown brought fossils collected at Mount Carmel to England. The fossils, reflecting a series of remains from the Stone Age Natufian culture, were discovered embedded in extremely hard limestone. McCown worked with his mentor, Keith, through the bitterly slow process of extracting the remains. Hrdlička, in the United States, regularly wrote to Keith inquiring about the team's progress and steadfastly applying pressure for the team to publish its findings in his *American Journal of Physical Anthropology*.

In January 1935, Keith wrote to Hrdlička optimistically assuring him that four people were working on the fossils and that the impending results were likely to add to the available literature on human history in the Middle East. He added, "They are strange folk the Carmelites, strange mixture: but the anatomical features of Neanderthal man predominate."[55] Hrdlička responded by assuring him, "I am glad to have your letter . . . and I am particularly happy to have the news that you are well and working on the remains from Palestine. It is a pity that such valuable things are so hard to get at and so imperfect but the results will, I hope, repay the drudgery." Hinting at the lack of reliable information available in the scholarship on prehistoric human remains, Hrdlička added, "I have heard recently that the French have discovered two or three skeletons of a similar nature—you probably know more about it that we do here."[56]

As the remains of more than thirteen individuals were painstakingly extracted from the limestone, it became clear that they possessed traits representing transitions between ancestral and modern humans.[57] Several years after the initial correspondence between Hrdlička and McCown about the Mount Carmel fossils, Hrdlička offered an entire issue of the *American Journal of Physical Anthropology* to McCown and Keith. McCown eventually utilized the results of his study in his doctoral dissertation. While his future research endeavors had only a modest influence in the field, he successfully trained a large number of younger scholars upon his return to the University of California, Berkeley, as a faculty member.[58] Historians continue to debate the place of McCown's discoveries within the broader timeline of human evolution.[59]

By the outset of the International Symposium on Early Man in 1937, however, it was already clear that scholars like McCown were experiencing transition. At the symposium itself, comparatively younger scholars like McCown and Frank H. H. Roberts, Jr., were joined by older figures such as Aleš Hrdlička, Eugène Dubois, and Robert Broom. Quite unlike earlier gatherings in North America, none of the thirty-six papers published in the symposium proceedings addressed the subject of race directly, instead focusing primarily on the study of human evolution through geology, climatology, anthropology, and archaeology. Although several papers reflected an ongoing interest in the archaeology and prehistory of North America, they generally explicated a global endeavor to understand the deep history of human evolution. Presented together with a small series of exhibitions in Philadelphia, the papers relied far more on Old World fossilized remains and artifacts than on the more recent remains found in museum collections in the United States.[60] Several papers stood out as markedly demonstrating a growing interest in questions brought about by people discovering new artifacts in North America. Taken together, these artifacts pointed to a much deeper or more ancient history of humankind's occupation of the Americas than was previously known.[61]

THE PEOPLING OF THE AMERICAS

In 1926, a discovery near Folsom, New Mexico, shifted the discourse about the arrival of modern humans in the Americas. One archaeologist, reflecting thirty years later, described the find as having "marked the beginning of a whole new field of archaeological research."[62] Similar artifacts began to emerge throughout North America. Just three years after the initial discovery of the Folsom culture, a series of beautifully crafted, fluted projectile points were discovered in Clovis, New Mexico, recognized as proving that an even older culture existed in North America than was previously assumed.

Before the new discoveries in North America, headlined by the unearthing of new stone tools in places like Folsom and Clovis, evidence for the early occupation of North America was remarkably limited. Scientists at the Smithsonian Institution—in particular, Hrdlička and Holmes—recommended that scholars proceed with extreme caution in pushing back the date of human arrival in the Americas.[63] Conversely, other scholars viewed new stone tool discoveries as validating their existing ideas regarding humans' early arrival in North America. Some even began to actively seek ancient skeletal remains to support their claims.[64]

In 1927, J. D. Figgins, the director of the Colorado Museum of Natural History, described the available skeletal or archaeological evidence surrounding the question of North American occupation as "exceedingly meager." In an article synthesizing the available materials, he elaborated by saying that the evidence was "far too scant to make possible intelligent comparisons and safely arrive at definite conclusions."[65] Nevertheless, scholars continued to combine evidence from archaeology, physical anthropology, and geology to create an increasingly more complete picture illustrating humans' early occupation of North America. While new archaeological evidence emerged in this era, human remains reflecting Paleo-Indian populations remained extremely limited before World War II.[66]

In the wake of discoveries at Clovis and Folsom, several scholars turned their attention to Alaska, hoping to expand the "meager" evidence illuminating the process of the peopling of the Americas. For certain anthropologists, the Alaskan Eskimo preserved numerous secrets regarding the arrival of modern humans in North America, and the discovery of ancient human remains in the region was thought to provide an opportunity to explore certain questions in more depth. The bodies of the recent dead were evaluated side by side with ancient skeletons in the hope of making some effective anatomical comparison. Following a research trip to Alaska in 1926, Hrdlička suggested that a new curator, Henry Bascom Collins, travel north to research and collect human remains and archaeological material. Collins found Alaska to be largely unexplored and unknown following the work of the nineteenth-century naturalist Edward W. Nelson. Collins stated in an oral history, "There had been no commercial exploitation of the Bering Sea areas whatsoever. [It was] a backwash. The people had not been described or visited almost since Nelson's time forty-five years earlier. They were the most primitive Eskimos anywhere in the Arctic."[67] During Collins's first year in Alaska, in 1927, he collected some skeletal material and acquired bodily measurements of the people in the region. Collins experienced a mixed record of success over the course of his first three field seasons. He managed to gather a small but unique skeletal collection, yet his ideas surrounding the region's prehistory developed only slowly, based on meager bits of archaeological evidence collected over shortened field seasons.[68] Eventually, Collins's work shifted away from physical anthropology, and he became one of the leading archaeologists to examine Alaskan and Canadian Arctic prehistory. His early career trajectory, however, reflected the interests of both Hrdlička and the broader anthropological community in the era, seeking answers to the question of the arrival of modern humans in the Americas through the study of human remains discovered in the Arctic.

In 1935, J. D. Figgins, the same archaeologist who described available ancient skeletal evidence in North America as meager, caused a stir by claiming that a recent discovery of a skull linked to

Folsom sites was the impetus for naming a new human ancestor. After receiving the bones from a local discoverer, Figgins brought them back to the Colorado Museum of Natural History. The remains included several parts of the skeleton, but scientists were mainly interested in the skull and the antiquity of the remains. After closely examining the shape of the skull, Figgins concluded, "the individual occupied a position intermediate between those of the primitive types of Europe and those of the modern races."[69] Figgins went so far as to declare the discovery a new species, *Homo novusmundus*. After traveling to Denver, Frank H. H. Roberts, Jr., of the Bureau of American Ethnology convinced Figgins to allow him to take the skull back to the Smithsonian. Roberts showed the remains to Hrdlička, Stewart, Collins, and the archaeologist F. M. Setzler. The scientists were in agreement that while the skull may have exhibited some "inferior" features, it displayed nothing to convince them that it was either "very primitive or un-American." Scientists at the Smithsonian dismissed the claim that the skull represented a new species as being "fanciful and wholly unjustified."[70]

Although the scientists examining the remains were interested in the skeleton's racial origin, their primary concern was locating the remains historically. The skeleton's potential value, they believed, was in addressing a burning question about human prehistory—even if it meant declaring that human ancestors had migrated to the Americans only after the arrival of *Homo sapiens*. If remains were to be found representing new species of humankind, they would not be found in the Americas. But this did not settle the question of when people first arrived in the Americas. In fact, efforts to answer the question of the antiquity of humans in the Americas only increased as the century progressed, following continued growth in both archaeology and physical anthropology. Despite the centrality of ancient skeletal remains in attempts to answer these important questions, ethical issues regarding the practice of collecting and studying the dead in the Americas were starting to emerge.

EMERGING INDIGENOUS PUSHBACK

In most cases, American Indians were adamant in their opposition to graves being desecrated by scientists who hoped to acquire skeletal specimens. Early collectors sometimes noted feelings of danger and guilt in collecting from graves in the name of science.[71] On the other hand, collectors who were interested in gathering human remains had not always felt obligated to keep secret their intent to remove bones from burial sites. By the late 1920s, when the Smithsonian archaeologist Henry Collins described his method for removing skeletal remains from ancient gravesites in Alaska, he noted that four local Eskimos even helped him dig, continually uncovering new graves. Once remains were found, Collins and another archaeologist would remove the bones from the ground and pack them to be shipped, apparently encountering no resistance from the indigenous workers whom they hired.[72]

Reaction to the removal of human remains from a gravesite or an accidental burial was anything but uniform, and archaeologists had written and spoken of the inherent dangers of collecting bodies for decades. During the 1930s, however, the pushback of indigenous peoples against collecting human remains started to become visibly apparent in internal museum correspondence. In 1933, even after a proposed Smithsonian expedition secured additional funds from the Museum of the American Indian in New York City, Smithsonian officials canceled a planned collecting expedition to the Aleutian Islands. Hrdlička was upset to the point of disgust at the rejection. He wrote in a letter, "one of the arguments raised against our fieldwork this year was that 'such trips irritate the people.'"[73] George Heye, the director of the Museum of the American Indian, responded to Hrdlička's letter, "I am truly sorry anybody in the world can be irritated by scientific work."⁷⁴ Many scholars of the period who were concerned with collecting bodies for science simply could not imagine weighing the religious, spiritual, or cultural rights of indigenous peoples against the goals of science.

Collecting human bodies, sometimes even taking remains from cemetery sites and recent battlefields, "irritated" indigenous communities for obvious reasons. Removing ancestral remains without permission upset the religious and moral sensibilities of many different cultures around the globe. This sort of cultural conflict, frequently pitting white scientists who hoped to gather remains for science against indigenous people who hoped to protect their sacred gravesites, quickly became apparent to those who were supervising the growth of the Army Medical Museum following the Civil War. Interactions of these kinds were typically marked by an imbalance of power. Later, tensions became increasingly clear, and obtaining remains in North America became an ever-greater challenge for museum scientists. The ethical ramifications resulting from these interactions are still being dealt with in museums today.

Indigenous resistance to the collection of human remains was starting to occur not just in North America. One year before Hrdlička's rebuffed attempt to travel to the Aleutians to collect remains, another anthropologist, Melville J. Herskovits, wrote to him describing the difficulty in obtaining skeletal material in Africa: "I am afraid I must disappoint you and tell you that I did not bring any skeletal material home from West Africa. I saw some marvelous collections of skulls that I itched to get but they were all in shrines. As a matter of fact, with the extent to which the ancestral cult is prevalent in West Africa, I seriously doubt whether skeletal material could be collected on the west coast without involving a general uprising of the native population."⁷⁵

Indigenous resistance increasingly became better organized and resulted in new legal protections. The first recorded successful repatriation effort in the United States took place in 1938, when a sacred bundle was returned to the Hidatsas.[76] Growing calls to return sacred objects and human remains came to dominate public discourse surrounding museum collections during the second half of the twentieth century. These ethical rejections, coupled with the lingering notion that the general public was repulsed by displays of human remains, removed all but a few displays of human remains from museums in the United States throughout this period. This removal from public exhibits to behind-the-scenes storage, however, did little to fully deal with the ethical problems inherent to the story of building these collections over time. By the conclusion of the Second World War, mummified remains from Egypt and Peru continued to draw museum visitors, but other types of mummified and skeletal remains, especially those from North America, were mostly removed from exhibit.

T. DALE STEWART

Born in 1901 in Delta, Pennsylvania, T. Dale Stewart began working for the Smithsonian in 1924, taking time away from the museum to earn an A.B. from George Washington University and an M.D. from Johns Hopkins University. Upon the receipt of his medical degree, Stewart was promoted from an aide to an assistant curator of physical anthropology at the museum. He rose through the ranks to become head curator in the anthropology department by 1961 and served at the director of the National Museum of Natural History from 1962 to 1965. Stewart was professional and at times might have been read by others as outwardly formal. But he also possessed an affable and engaging personality. Over the course of his career, he brought in a number of students and encouraged visiting scholars to study the Smithsonian's collections, a marked shift from the guarded curation practiced by Hrdlička. Although Hrdlička largely overshadowed Stewart in total output, Stewart's contrasting personality encouraged new innovation and scholarship applied toward human remains collections. Hrdlička had been cold, dogged, and sharp-tongued. Stewart, on the other hand, was unusually thoughtful, and his critiques were rarely read as direct or personal attacks.

Stewart became a prominent, and slightly senior, member of a new cohort of physical anthropologist in the United States who came to change the field. These scholars included Marshall T. Newman (1911–1994), Sherwood Washburn (1911–2000), and John Lawrence Angel (1915–1986). While these scientists remained interested in certain questions surrounding race, much interest moved from earlier racial classification schemes to questions about the historical development of populations or races in different regions. Certain scholars within this cohort began abandoning questions of race altogether, instead turning their attention entirely to the development of human anatomy and evolution. Stewart officially retired from the Smithsonian in 1971, but he remained active in anthropology for years. Unlike scholars who benefited from the legacy building of graduate students who worked under their supervision, Stewart's influence in the field of physical anthropology came from his position of leadership supervising the human remains collections at the Smithsonian—which continued to house the largest physical anthropology collection in the nation. This responsibility came along with continued involvement in academic journals and professional organizations.[77]

Stewart worked directly under Hrdlička as a temporary assistant before being encouraged by his supervisor to pursue a medical degree.[78] Before Stewart entered medical school, Hrdlička sent him to collect skeletal materials in Nunivak Island in Alaska. Stewart remembered the summer in his oral history: "We would make treks across the island to abandoned villages and collect the dead around those villages and any cultural objects that we could find. It was a remarkable summer there."[79] Stewart's fieldwork in

Alaska represents a continuation of an established Smithsonian tradition of collecting human skeletal remains in Alaska. Upon his arrival at medical school, Stewart began studying under doctors who were also interested in physical anthropology.[80]

Stewart, unlike many others in the field of physical anthropology, worked well with his mentor, Hrdlička. Stewart recalled of Hrdlička, "I got along with him all right simply because I didn't try to counter him. If he wanted things done certain ways I attempted to do them the way he wanted but would try insidiously to suggest ways of improving it."[81] Hrdlička eventually grew to trust Stewart and assisted him by organizing research projects and fieldwork and arranging a leadership role for him at the *American Journal of Physical Anthropology*. Stewart stepped into these responsibilities capably, and following Hrdlička's death in 1943, he became a prominent figure in physical anthropology in the United States.

RACE, THE PAST, HUMAN BLOOD, AND HAIR

Just as human remains collections modernized and grew in size and scope in museums, researchers in physical anthropology and medicine started to turn some attention away from human bones and toward soft tissue, blood, and even human hair. Years earlier, researchers at institutions like the Army Medical Museum had shifted their emphasis from the "comparative anatomy" in racial classification toward studying communicable disease. This involved studying and collecting soft tissue, which was thought to provide information that was more pragmatic to the medical researcher who was interested in disease. Despite the apparent utility of these types of human remains, they were much more challenging to preserve than were human bones, which typically sat untreated, for long periods, in a bone room with only rudimentary climate controls. While the Army Medical Museum worked in the late nineteenth century to purge its collections aimed at studies in racial classification, the

Smithsonian and numerous other museums interested in natural history continued well into the twentieth century to maintain an active program in collecting remains from around the globe for the purpose of racial classification.

As researchers in medicine became increasingly concerned with soft-tissue samples for the purpose of understanding disease, new work in physical anthropology started to focus on the same tissues, blood, and hair in order to understand lingering theories of racial classification. Advances in genetic research and the discovery of the double helix eventually opened up countless new avenues for this sort of research, but even before the start of the Second World War, medical museums advanced efforts to collect and catalogue hair and soft-tissue samples. At the same time, many scholars were becoming increasingly concerned with the ancient past. The hair and skin tissue of mummified remains, it was soon discovered, provided a particularly valuable resource with which scholars could learn about the distant past.[82] Taking lessons learned from the study of soft-tissue samples related to contemporary health issues, scientists began to collect ancient samples, examining tiny samples under microscopes to learn more about ancient bodies.

Some of the earliest signs of these shifts arrived through the work of an anatomist and physical anthropologist named Mildred Trotter (1899–1991). After training at Mount Holyoke College, Washington University in Saint Louis, and Oxford, Trotter returned to Washington University, where she joined the faculty of the medical school. Trotter proved early on in her career to be a talented researcher, and she split her time between studying hair samples and bones. She found working with human bones to be more interesting than work with human hair, but she quickly became an expert on the latter subject and published numerous papers comparing the hair of various races.[83] Trotter attempted to construct a typological understanding of modern human hair; however, she also became interested in available hair samples from older, mummified remains.

Museums in the United States had long maintained an interest in collecting mummies, and Trotter found many curators quick to comply with requests for mummified tissue samples. Now understanding the potential value for new methods in comparative hair research, she eventually worked to examine numerous hair samples from human remains collections around the globe.

Trotter was careful to note that mummified hair had been studied before her own scholarship. Researchers from the Historical Society of Colorado, for example, became interested in comparatively studying hair in the 1930s, and their research comparing both contemporary and historic tribes included studying the hair of ancient Puebloan mummies.[84] Trotter's work built on earlier studies through advancing the overall scope as well as the technological and methodological sophistication. Embracing both the microscope and statistical methodologies, she worked to place the development of hair alongside the data available for the development of the skeleton. Between 1922 and 1973, Trotter published over twenty articles on hair. One article, published in 1943, utilized ten scalp and hair samples from mummies housed in museums in Peru—T. Dale Stewart had personally collected the samples for Trotter from the Smithsonian's bone rooms. Trotter then carefully examined the fragile samples before describing them in the *American Journal of Physical Anthropology*. In the article, Trotter notes that the mummified hair subjected to study by scholars working in the United States had both lightened in color and dehydrated since death. Trotter concluded that while the hair samples of the Peruvian mummies she examined varied widely, they were similar enough to the hair samples of other American Indians, both living and ancient, to maintain existing racial classification schemes.[85] Trotter's studies, while continuing to emphasize schemes of racial classification well into the 1940s, began to blend a pure interest in the study of contemporary race with a more historicized study of humankind.

By the 1940s, William C. Boyd, an immunochemist, argued that blood type should be central to physical anthropology. Boyd studied the blood of living humans as well as trace samples from mummified bodies in museum collections.[86] Like many of his contemporaries, Boyd was growing interested in what historic and prehistoric populations might teach scholars about living populations. Scholars of this period were also becoming increasingly interested in understanding how living populations might hold keys to understanding prehistory.

Though some scholars found information emerging about blood type to be useful to the study of race, many anthropologists were skeptical of the manner in which finds were presented. In writing to Boyd, T. D. Stewart explained his mentor's position: "Dr. Hrdlička is doubtful about the value of some of the blood group data because of the factors of race and mixture and different serological techniques."[87] Boyd acknowledged these difficulties, explaining that the existing research on blood type and race was "of very unequal merit."[88] Hrdlička, on the other hand, found Boyd's work on blood type to be compelling and was willing to share mummified tissue samples from Egypt and the American Southwest, as well as an assortment of other soft tissues in the Smithsonian's collections, such as human brains and scalps.[89]

Nevertheless, museums were curious as to what sort of information might be learned from blood-type studies of mummified tissues. When samples of human skin were first solicited by Boyd for his studies on ancient blood types, the Boston Museum of Art and the Smithsonian both sent samples from ancient Egyptian mummies. Additionally, the samples from American Indian mummies from the American Southwest (referred to as the "Basket Makers") were first in line for study. Boyd wrote in a letter to Hrdlička, "It seems to me that these results are of some interest, especially that with the Basket Maker. It would be of interest to examine other Basket Makers to see if they too, differ from the other Indians."⁹⁰

From the middle of the 1930s into the 1940s, medical doctors and anthropologists interested in studying human hair and human blood crafted an intersection between the study of racial classification and the study of history that paralleled the construction of exhibitions on similar subjects. Though their research does not represent a wholesale, overnight shift from the study of race to the study of prehistory, scholars began to address the subject of prehistory and human evolution more fervently, and their work proved consequential to others in the field. The availability of mummified tissues stored in bone rooms in the United States from Peru, Egypt, and the American Southwest reflects the history of collecting practices from the middle of the nineteenth century to the middle third of the twentieth century. The history of collecting patterns, therefore, had a direct influence on the development of a more modern battery of scientific tests created for blood, tissue, and hair samples from ancient remains. The results of these studies had influence on understandings of both race and human history, but the shift from a pure interest in racial science to studies focusing on a deeper history was increasingly apparent.

HUMAN REMAINS COLLECTIONS, THE UNIVERSITY, AND THE MUSEUM

In 1975, in an article examining the recent expansion in physical anthropology, T. D. Stewart made the argument that, "unlike the rest of anthropology, physical anthropology moved from a museum phase into an academic phase around 1940 rather than in 1900."⁹¹ Stewart began his article by noting that the professional field of physical anthropology was quite small in the United States in the 1920s. By the middle of the 1970s, as Stewart described, the field was still based heavily on the direct examination of human remains and was in many respects flourishing. Stewart attributed this largely to one individual, his mentor, Hrdlička. When Stewart arrived at the Smithsonian, Hrdlička was a giant in his field, having built an expansive global collection. When Hrdlička died in 1943, however, the bone collector's stature in the field had diminished. His ideas failed to evolve even as others brought to light new evidence. Despite these changes, still only a few physical anthropology courses were even available in universities in the United States, and the few that were available were centered on small clusters of specialists.

Stewart's narrative of the history of physical anthropology begins with his own arrival at the Smithsonian, working as Hrdlička's assistant, in 1925—the same year Hrdlička ascended to the position of president of the American Anthropological Association (AAA). Stewart notes that the meeting of the AAA that same year featured forty-one total papers and that sixteen of those papers—or nearly 40 percent—were related to physical anthropology. The national meeting of anthropologists featured papers on topics of physical anthropology read by geneticists, anatomists, and individuals like Melville J. Herskovits and Franz Boas, both of whom are largely remembered as cultural anthropologists.

Despite Stewart's claim for a continuance of museum-based physical anthropology, a shift toward the university had already begun in the 1920s. By 1925, Stewart notes, of the eighteen associate editors of the *American Journal of Physical Anthropology*, five were from medical schools, five were anthropology faculty at universities, four were from endowed research organizations, and four were from museums (and two of these were university-based museums).[92] Stewart was responding to a portrait of museum anthropology drawn by Clark Wissler at the American Museum of Natural History (AMNH). Wissler had a specific interest in the history of museum anthropology in the United States and had been a curator in New York for some time. Wissler based his claims on the total number of acquisitions by his own anthropology department at the AMNH, rather than physical anthropology collections in particular.[93]

Others in the anthropological community echoed Stewart's argument that physical anthropology struggled to take hold within anthropology departments in universities before the Second World War. In 1934, Edgar Hewett made an address in which he stated his belief that physical anthropology remained outside the realm of university-based anthropology, after which he received a critical reaction from Hrdlička. Hewett wrote in response, "I am confronted with the obligation of building a department of the science of man, in which Physical Anthropology should be a vital factor. With that in view, I have for the last two or three years been studying the Physical Anthropology work done in our American universities, and I have reached the conviction that for the most part it is puerile and profitless. I have not found a single university in which there is a clear statement of the inherent values of Physical Anthropology, or a clean cut statement of its application to the problems of modern life."[94]

Hewett, in working with Hrdlička on the *Science of Man* building for the 1915 Panama-California International Exhibition, had personally hoped that the exhibits would spark universities to include physical anthropologists within their expanding anthropology departments. Hewett lamented nearly twenty years later, "How unfortunate that we could not have produced that convincing statement of the science of man in connection with a great educational institution."[95] Hrdlička, in responding to Hewett's lament, argued that failings in physical anthropology were not due to their own efforts in exhibition. Instead, they were due to others who misrepresented the science in public settings. Additionally, Hrdlička believed that "the biblical tenets are still very strong—and that even with many educated people, which results in many still looking on physical anthropology [as] something dangerous and even subversive." As to the failing of physical anthropology in the university, Hrdlička explained his belief that "the college people hesitate to give it due chance, both for personal reasons and for reasons of policy." Hrdlička continued his letter critiquing the address given by Hewett by underscoring the importance of training the next generation of scholars to work with human bodies: "Under [existing] conditions no one can wonder that the men who represent [physical anthropology] are so frequently amateurs, who are underinstructed, often biased, and following trivial if not destructive tendencies." Nevertheless, Hrdlička predicted, "A hundred years hence physical anthropology in this country shall have become thoroughly established."[96]

Confiding in other scholars just a few years later, however, Hrdlička's tone did not seem quite as confident. In a dense letter written to Sir Arthur Keith, still considered a preeminent figure, Hrdlička explained his view on the state of American anthropology. He echoed his idea that the other branches in anthropology shunned physical anthropology. He added, "In addition there has developed during the last 15 years a curious condition from which probably you do not suffer. This is the fact that a good many of our anatomists, under peculiar influences, lack medical preparation. They are just Ph.D.'s, and thus handicapped in relation to Anthropology."[97] Hrdlička's estimation—that an increasing number of individuals concerned with human anatomy, evolution, human prehistory, and race were coming out of anthropology departments, as opposed to medical schools—was correct. By the time Hrdlička died in 1943, the use of human remains for research and display had largely shifted away from scholars trained in medicine toward those who were completing doctoral work in anthropology.

In T. D. Stewart's conclusion to his later analysis regarding the history of physical anthropology, he argued that bone acquisitions at the Smithsonian ran somewhat counter to existing notions about a "museum period." Hrdlička arrived at the Smithsonian only in

1900, he reminds the reader, and his vast collection was mainly acquired by the museum during Hrdlička's active years of collecting—between 1903 and 1943. Stewart writes, "Taking this into account, along with the events between 1930 and 1950 enumerated above, I am inclined to advance the dividing point between physical anthropology's museum and academic periods to the neighborhood of 1940."[98] Although it is impossible to truly pinpoint the transition of physical anthropology from a discipline based in museums to an academic field more firmly ensconced at universities, it was Stewart's impression that museum bone collections, built over the previous half century, continued to be central to physical anthropology in the United States until at least the Second World War.

While Stewart was complimentary of Hrdlička's legacy in his writings near the time of his mentor's death, others in the scholarly community were not so complimentary of the Czech-born scholar's influence. Writing many years later, Sherwood Washburn, a British-born scholar who worked to modernize the field of physical anthropology through comparative anatomy, argued that Hrdlička's influence has seriously waned in the academic community toward the end of his life. Washburn wrote of Hrdlička, "he very nearly killed physical anthropology." He continued, "By the time I was in college he was regarded as an old, disagreeable, fool."[99] Hrdlička had always been challenging to work with, and many of his unchanging ideas were considered outdated by the 1930s. Nevertheless, Hrdlička held sway over one the largest and most complete collections of human remains in the country. Further, his influence had guided numerous scholars across the country, and he helped construct the framework for building and organizing bone rooms around the world. Significantly, in his *New York Times* obituary, Hrdlička was largely eulogized as an important scholar of human evolution. The obituary reads, "Forty years of measuring convinced Hrdlička that man sprang not from some anthropoid ape, as Darwin postulated, but from some vanished creature more human." Despite some fairly egregiously broad and even inaccurate descriptions of the history of science and studies in evolution, Hrdlička's obituary is telling in that it primarily emphasized the aspects of his research focusing on humanity's seemingly distant and mysterious past. "In these bones," the obituary reads, "the secret of man's problematical descent was to be read." The memorial concludes, "Time alone can tell how much of Hrdlička's case will stand. But there is no uncertainty about his place in physical anthropology. To fill that place is impossible. His successor must of necessity strike out for himself. Hrdlička could not have exhausted the possibilities in the mountain of material he collected, but he did exhaust them so far as our present knowledge of human evolution is concerned."[100]

The *New York Times*'s conclusion—that human remains housed in bone room would continue to inform our understanding of human evolution—was a powerful line of thought that persisted into the second half of the twentieth century. To this day, some scientists working with medical and natural history museum bone collections are reticent to repatriate or rebury ancient remains, as they continue to yield new discoveries with technological advances.

Much of the available literature on the history of museum anthropology upholds the notion of a "museum period" or "museum age" in the United States. Such a period, however, has always proven difficult to define, and the way in which physical anthropology fits within this story is complex. Scholars hoped that the research and display of both living humans and human remains would inspire the addition of physical anthropology to the university in the United States, but until the Second World War, such an occurrence largely failed to take place. At the very least, the history of collecting human remains at major museums should complicate the existing notion of a museum period for museum anthropology in the United States.

PHYSICAL ANTHROPOLOGY—
STILL ANTHROPOLOGY?

Although physical anthropology remained one of the four major subfields of the discipline of anthropology, some people felt the field was becoming increasingly isolated from a larger anthropological community. Physical anthropologists increasingly relied on their own journals to disseminate new ideas and announce the discovery and acquisition of new specimens. In 1933, when a position opened up for the anthropology section of the National Academy of Sciences, Hrdlička lobbied that the position be filled by a physical anthropologist. He noted the cooperation between physical anthropology and other disciplines, reporting that the field was "gradually forging its way to our Universities."[101] Nevertheless, there remained a perceived lack of disciplinary representation in scientific and anthropological organizations. While other branches of anthropology largely moved from the museum to the university, physical anthropology remained somewhat embedded in the museum, married to the collections of human remains that were so central to studies since the Civil War. Hrdlička worried that his field might be left behind by the other subdisciplines in anthropology.[102] Despite his unwavering belief that physical anthropology was destined to grow, Hrdlička wrote contradicting opinions on the future of physical anthropology in universities in the United States.

Some observers considered the 1930s as a decade with little progress for physical anthropology. A handful even viewed studying museum bone collections as old-fashioned, instead traveling abroad to uncover and study fossilized remains in places like Africa and Europe. As cultural anthropologists moved away from object-based epistemology, physical anthropologists were still immersed in studies of human remains. Melville J. Herskovits, the prominent anthropologist who built his career researching in Africa, hoped to reunite the subfields of anthropology, which he believed to be drifting apart. In a letter applauding T. Dale Stewart's efforts as the editor of the *American Journal of Physical Anthropology*, he wrote, "It will be a pleasure to cooperate with you, either in doing an occasional review or perhaps, if it is possible these busy days and you be interested, in sending you a paper incorporating some observations I have been making over the past few years about the unity of anthropology, with particular reference to the problem of whether both physical and cultural anthropology cannot be mobilized to make for a better analysis of certain problems than either can achieve alone."[103] Physical anthropology, in Herskovits's view, had drifted into "doldrums." Anthropologists, he argued, might be encouraged to address particular questions through a more unified, or interdisciplinary, approach.

Herskovits was prompted to write to Stewart upon reading a recent editorial that appeared in the *American Journal of Physical Anthropology*. The editorial, which appeared on the opening pages in a new series for the journal, announced the transition of the editorship from Hrdlička to Stewart: "Looking back over the 29 volumes of the Journal that have appeared since 1918, Doctor Hrdlička's foresight and courage in initiating single handedly a publication in such an undeveloped field as physical anthropology, and especially during a world crisis, seem monumental. Not many visualized the vast materials waiting to be studied or realized the need for a medium in which to record the work waiting to be done."[104] The editorial explained the vast influence Hrdlička wielded in shaping the early development of physical anthropology as a field. As an editor and as "one of the few full-time physical anthropologists" in the entire nation, he influenced how human remains came to be catalogued, displayed, measured, and interpreted in museum collections.[105] Once the shift of editorship was finalized, Stewart publicly recognized the opportunity to impose "liberalizing changes" in the field.[106] Reflecting broader changes in the discipline of physical anthropology, the new journal editors hoped to diversify the scholarship, moving away from the skull measuring of craniometry (which did remain of special interest) toward a more complete study of the human body.

Changes in physical anthropology were not, however, immediately recognized in the broader anthropological community. In 1945, when the journal *American Anthropologist* published a bibliography of recent publications in anthropology, the list included citations of recent work in ethnology and archaeology but not in physical anthropology. Stewart wrote to the editor of the journal, J. Alden Mason, who replied that the omission was merely an oversight: "Physical anthropology is the phase of anthropology in which I am least interested and informed. I had never noticed the omission."[107] Stewart, for his part, conceded that "everyone realizes that the *American Anthropologist* has specialized in the field of ethnology."[108] Nevertheless, the omission can be taken as clear evidence of physical anthropology's somewhat wayward and disconnected position in the United States at the outset of World War II. The field's influence appeared to be waning both within the larger field of anthropology and in American intellectual and cultural life more generally.

That same year, as the Second World War ended, Sherwood Washburn wrote to William Duncan Strong expressing similar concerns about the fate of physical anthropology. Washburn, an anatomist, primatologist, and specialist in human evolution, wrote to Strong hoping to convince him to join the American Association of Physical Anthropologists. Strong, an anthropologist and archaeologist who never seriously studied human remains, seemed an unlikely candidate to join an academic association of physical anthropologists. Washburn assured him, "I really think that the Association is picking up a lot." He added, "We've lots of anatomists in it. What we need now are anthropologists."[109] Despite the best efforts of scholars, anthropology was becoming an increasingly divided field—split into communities of specialists. Physical anthropology, long centered on the research and display of collections of human remains in museums and medical schools, struggled to solidify an identity in the middle of the twentieth century. The era of rapid collection, publication, and display of human remains in museums, fairs, and medical schools that had consumed hundreds of anatomists, anthropologists, and medical doctors in the United States was ending. A new era in the study of human evolution, however, was just beginning.

. . .

In 1936, the New Deal made possible a new partnership between the Smithsonian and the Department of Interior. Together, they created a new radio program boldly called *The World Is Yours*. The program's reach grew slowly, but after two years, the show boomed over the airwaves through nearly sixty radio stations across the country. By the end of its second year, the program received nearly a quarter of a million letters from listeners. The show officially began in July, and in mid-August, it aired an episode titled "The Story of Man in America." In November, an episode titled "The Evolution of Life" hit airwaves. Following that episode, the public listened to a program called "Early Man." What the program lacked in scientific detail and accuracy (even for the era in which it appeared) it attempted to make up for through dramatic language and vivid description. Despite deficiencies, the program brought the subject of human evolution and the study of fossilized remains to massive audiences. The radio program explained to listeners, "Within the last 50 years, through fortunate discoveries, science has been enriched by a number of very ancient fossils which cannot be identified positively either as human or precursor of human." It continued, "The riddle of man's past is unraveled by scientists just like any other major natural riddle, by starting with known facts in the present, and deducing the nature of the unknown from well authenticated remains of the past."[110]

During the first two years of the radio program, which Smithsonian leaders believed to fit neatly into their mission of disseminating knowledge, the subject of race was notably absent. Instead of hearing ideas about racial typology, listeners heard about human evolution and prehistory.[111] When listeners *were* finally presented a radio program on the topic of race, the show examined the concept of "racial equality," rather than the supposedly scientific study of racial classification, which continued to fall out of favor. Although physical anthropologists spent several generations researching and displaying collections of human remains for the purposes of comparative anatomy or racial classification, the communities they helped create became increasingly focused on human evolution. Certainly, an interest in comparative racial studies lingered for those who were concerned with researching and displaying human remains, but the major concentration of the discipline started to shift. Those who entered the bone rooms possessed different goals than their predecessors had, and the conclusions they drew from studying the dead were vastly different from those of similar scholars only a few generations before. Not only were discussions of racial typology disappearing from the public discourse of anthropology, but also artistic representation of ideas drawn from the study of human remains largely replaced the display of actual human bones in museum exhibitions in the United States. The hundreds of thousands of human remains held at major museums in the United States were largely relegated back to bone rooms, behind the scenes.

In 1943, Hrdlička reflected in a brief unpublished manuscript, just a short time before his death, on the history of physical anthropology in the United States. At the conclusion of his essay, he turned his attention to the future of the discipline: "A vast amount of [the work of physical anthropology] remains still to be done, both on skeletal materials and on the living."[112] Though Hrdlička was enthused by certain developments among the younger generation of physical anthropologists, he was reticent to see other developments introduced into the field that he had helped create. Hrdlička certainly could not have imagined how studies on human remains were to change over the next half century.

The broad intellectual transition experienced in physical anthropology was reflected in the depiction of human remains in exhibitions. In 1915, at the Panama-California Exposition in San Diego, the *Science of Man* building introduced basic concepts of human evolution and human prehistory early in the exhibit, but the majority of the displays focused on ideas surrounding racial classification. By 1933, the Field Museum of Natural History opened a pair of exhibitions placing racial classification and human prehistory on nearly equal footing (at least in the size and scope of new exhibit spaces). By 1941, when the Smithsonian Institution opened new exhibits introducing the public to the broad scope of the many branches of the institution, it featured a display on physical anthropology that blended ideas about human evolution and comparative racial studies. All three exhibitions leveraged the use of human remains and reproductions of the human body to teach the public about shifting ideas in the scientific community. Without a doubt, the promotion of racial classification in museum exhibitions continued through the conclusion of the Second World War, but the major emphasis in most displays experienced a shift.

In a letter written many years after the Second World War, Sherwood Washburn, a prominent physical anthropologist, anatomist, and primatologist from the University of California, Berkeley, wrote that the very term *physical anthropology* was losing its resonance. He wrote simply, "Basically very few people are interested in physical anthropology. Human evolution is the area of interest." As opposed to the declining value of the discipline of physical anthropology, the term *human evolution*, he argued, "is a phrase that communicates."[113] Washburn himself joined numerous anthropologists in the outright denial of race as a viable concept in the second half of the twentieth century.[114] As scholars continued to leverage this growing interest in human evolution, their work was bolstered by a series of high-profile discoveries in the field of paleoanthropology. Though many continued to study remains stored in museums throughout the United States, the emphasis of their studies had clearly shifted. Only a few years before Washburn wrote his letter, another scholar examining the history of physical anthropology argued simply, "Human skeletal biology is moribund because of its long history of abuse by racial typologists and its largely descriptive nature."[115] Due almost exclusively to the abuse of racial science, bone rooms sat silent, the argument went. The reality, of course, was more complex. Indigenous activism, the decline of physical anthropology within the broader field of anthropology, new discoveries in paleoanthropology, and a generational change all played a part in the generally weakening significance of bone rooms in museums in the United States immediately before World War II.

Major concepts in human evolution also underwent a shift that defined research and display over the course of the twentieth century. In an introductory textbook published in 1945, M. F. Ashley Montague noted, "we see that earlier notions of a linear evolution of man, conceptions which held that man progressively advanced in a straight line from an ape-like stage toward the stage of *Homo sapiens* were too simplified."[116] Exhibitions created before the war—many of which directly linked race to a linear conceptualization of human evolution—seemed hopelessly out of date in the decades that followed. Although not everyone was in agreement with Montague's progressive stance on race, the weight of the physical anthropological community gradually shifted in his direction.[117] Physical anthropology generally recovered as a field, following a gradual decline in the middle of the twentieth century, revived by new discoveries, new advances in science, and new generations of scientists. Furthermore, shifting the emphasis into seemingly more distant or abstract questions in history did not wipe away the ethical concerns inherent to these studies.

The bones collected from around the world stayed essentially the same, but the ideas surrounding them continued to change for the next half century, just as they had in the previous half century. By the end of the twentieth century, it became clear that those who were interested in studying human remains collections would finally be forced to wrestle with the realities of the legacy of scientific racism. Studies in racial classification continued to morph into related research that focused on ancestry and population, these areas of study often progressively working to subvert race as a concept. New scientific fields, coupled with new discoveries in anthropology, provided scholars interested in human prehistory and human evolution with growing bodies of evidence to debate. Most notably, a series of challenges to the ethics of the display and research of human remains grew from sporadic instances of indigenous resistance to a more unified, multicultural, and global discourse that changed the nature of the enterprise of collecting and displaying the human body.

EPILOGUE

There is nothing natural about systematically collecting and studying the dead. Through a complex cultural process and evolving assemblage of ideas, however, such a practice became reality. Museum bone rooms expanded as a result, filling shelves floor to ceiling with mummies and skeletons. The future of these bodies, their exact meaning and use, continues to be deeply divisive. In the decades following World War II, museums holding human remains collections faced shifting and frequently competing desires. The desire to learn more from skeletons occasionally faced off directly with efforts to rectify past wrongdoings, bringing about unforeseen challenges at the dawn of the twenty-first century. In the mid-1990s, over a century after the death of the lone Dakota man described in the opening pages of this book, his earthly remains were finally returned to the Sisseton-Wahpeton Dakota Nation for reburial. The bones were still cracked and broken in places, bearing decades-old scars, and the Smithsonian worked in close consultation with tribal representatives to determine how best to carefully disassemble bones wired together in years past. When the man's bones were first acquired, they worked as a stand-in for all Native American bodies. Exhibited in a museum gallery as such, the skeleton was later stored in bone rooms organized by racial typology. Repatriation and subsequent reburial again remade the meaning associated with the bones, ultimately working to reveal the true humanity and individuality behind a previously scientized specimen. Only through the process of bringing him home was the young man fully humanized as an individual.

Yet the repatriation of the Dakota man's remains also represented a distinct sea change in thinking about human remains collections in the United States that started in the late twentieth century.[1] A pair of new laws, the National Museum of the American Indian Act and the Native American Graves Protection and Repatriation Act (NAGPRA)—passed in 1989 and 1990, respectively—finally opened the door to returning ancestral human remains. The laws represented a small step forward in decolonizing centuries-long practices of acquiring museum collections. While the laws proved incomplete and in some places highly problematic and controversial, they nevertheless represented a new chapter in the treatment of museum collections of human remains in the United States.[2]

In 2009, a U.S. Senate oversight hearing revisited the NAGPRA law. The hearing opened with a statement from Congressman Nick J. Rahall: "The human remains that are at issue are the ancestors of Native Americans, many of them warriors killed in battle. They deserve the same respect that we give to the human remains of our warriors today."[3] Though the recommendations that came from the hearing still leave notable gaps in the ability of contemporary American Indian communities to reclaim their ancestors still housed in museums, they represented yet another incremental advance in attempting to rectify some of the many past wrongs inherent to the colonial anthropology enterprise. Nevertheless, while the dramatic story of recovering human remains from battlefields represented a clear, understandable breech in moral conduct to contemporary members of Congress, the law is limited in its scope to Native Americans, Alaskans, and Hawaiians. The scope of the current law blinds some commentators to the injustices inherent to what was, in fact, a *global* project to collect the dead for museums in the United States. The project to collect human skeletons stretched well beyond gathering bodies of American Indian warriors from battlefields. Indeed, war and massacre frequently served as opportunities to acquire collections, but museums acquired bones through an extensive array of opportunistic practices. Ethical treatment for human remains collection should extend across the legacy of the entire project to build bone rooms.

By the time emerging repatriation legislation became law in 1989 and 1990, the Smithsonian alone had amassed roughly 33,000 individual sets of human remains—of which about 19,250 were identified as Native American. The museum soon acknowledged 5,500 individual sets of remains as falling under the rubric of the new law, eligible for return to ancestral communities. To date, however, only 3,939 bodies from the Smithsonian's collection have been claimed by Native American communities and, ultimately, repatriated.[4] This story is as much about continuity as it is about change.

In 1991, just one year after the landmark passage of NAGPRA, the archaeologist and museum curator David Hurst Thomas pondered the implications of the law for museums in a short essay, "Repatriation: The Bitter End or a Fresh Beginning?" Thomas correctly predicted, "This new law will, without question, refashion the complexion of museum anthropology in the very near future."[5] Indeed, the law resulted in complex and often dynamic paradigm shifts for museums. For some, losing scientifically valuable specimens was a lamentable consequence of the law. But for many others, the passage of repatriation legislation represented novel opportunities to partially address past wrongs while also establishing important new connections between indigenous communities and museums. While occasionally resulting in heated debate, repatriation programs have proven, for the most part, to be successful steps forward for museums large and small across the country.

NEW SHELVES, NEW IDEAS

Today, the Smithsonian's biological anthropology collection would, in certain ways, be astonishingly unrecognizable to Aleš Hrdlička and his bone-gathering contemporaries. New generations of scholars following Hrdlička both built on his legacy and rejected it. The scholars who followed continued to move toward questions more firmly grounded in human origins, comparative anatomy, prehistory, paleopathology, and ancestry while also moving away from the more rigid project of assigning racial categories to human societies.

No longer is race the primary lens through which these skeletons are understood. Skeletons are now arranged and comprehended both geographically and with regard to historical age, with more complex methods in archaeology and physical anthropology informing ideas about ancestry and origins. Fresh ideas and approaches to science have resulted in radical but incomplete shifts, making physical anthropology now distinct from the methods utilized by earlier generations. Advanced science now generally eschews racially classificatory schemes, instead studying human evolution and ancestry. The growth of paleoanthropology following the Second World War primarily emerged from university campuses, where faculty drove the search for fossils abroad, rather than the museum curators who previously drove the practice of establishing collections in bone rooms. As postwar medicine shifted, so too did medical museums such as the Mütter Museum. The growing reliance on genetics to understand the body and human difference proved difficult to effectively collect, exhibit, and display. This was not only true in medical museums; similar ideas were difficult to display in natural history museums as well. Newly emergent international patrimony laws handicapped the ability of museums in the United States to gather fossils and ancient remains from abroad, and gradually scholars in the United States adapted to these new regulations, partnering with local officials to continue the work of searching for human ancestors. Some museum collections, including the Smithsonian's, continue to grow, as donations are occasionally acquisitioned if they are deemed an asset to the bone rooms.

Recently, many overflowing collections have been moved away from the National Mall. Skeletons and mummified remains in the Smithsonian's collections are now housed in pristine, white rows of enclosed, movable shelving units. The top-of-the-line new storage space is part of an existing, climate-controlled museum storage facility in Maryland, located less than ten miles away from the National Museum of Natural History (NMNH).[6] The novel storage facility, distanced from the NMNH building opened in 1910, offers some significant advantages to museum curators and other researchers. Tests increasingly draw DNA from ever more modestly sized human bone samples and other preserved tissues, with each new technology making the process less destructive to the original samples. Portable CT scanners can be wheeled into bone rooms to create highly detailed scans of human remains. Three-dimensional scanners and printers can replicate fossils and bones in detail that was never before possible with epoxy casts. Human remains can be studied in innovative ways never before imagined possible by previous generations of scientists. Despite these obvious advances in methods, many indigenous people remain skeptical as to the promise of this new science, especially in light of their historical treatment in studies involving non-Western bodies and human remains, now critiqued as "biocolonialism."[7]

Despite continuity in collecting and preserving human remains in museum bone rooms, new ideas about these remains have displaced some older theories that drove the growing practice of bone collecting as established in the mid- to late nineteenth century. The anthropologists at the museum today, partly in recognition of Hrdlička's complex legacy, approach problems connected to storing and caring for human bones with both new science and new ethical considerations. Museums now consult with the ancestors of those who are housed in collections and since 1990 are sometimes compelled to return specimens from federally recognized Native American tribes. Whereas Hrdlička was driven by a burning desire to amass the largest bone collection in the United States, others, like W. Montague Cobb at Howard University or the physicians affiliated with the Mütter Museum, were motivated by entirely different ideas, including the desire to prove that racial hierarchies were scientifically untenable. The Mütter Museum, too, has recently built entirely new storage rooms. A recent campaign to conserve the museum's collections—called "Save Our Skulls" by the museum—garnered widespread attention. Still thought to possess a great deal of value as educational tools, museums like the Mütter Museum maintain that these collections also hold promise for scientific study—though not always for the same reasons conceptualized by the original donors.

While the many advances apparent in the Smithsonian's recently completed storage facilities might have surprised the collectors described in this book, the scientists and amateur collectors prominent in this story might also have felt quite comfortable in this glistening space, where an enormous collection of almost thirty thousand humans still lie silently. It is, in some ways, a natural extension of the project they developed and expanded during the era examined in this book. In other ways, this would appear to be an entirely novel enterprise. Recent discoveries stemming from new science like radiometric dating, advanced understanding of human genetics, evolution, and more finely tuned archaeological methods have revolutionized the study of the dead in museum spaces. Emergent engagement and consultation practices with indigenous communities, especially Native American communities following the passage of NAGPRA, would also be almost completely foreign to collectors who—in previous generations—held little regard for the objections to the gathering and study of ancestral remains.

Although new laws and ethics resulted in thousands of repatriations and reburials from museums both large and small, the law applies only to federally recognized tribes in the United States, leaving the bones of federally unrecognized or "culturally unidentifiable" remains on museum shelves. As the story in the preceding pages suggests, the history of collecting, researching, and exhibiting human remains at museums and world's fairs in the nineteenth and early twentieth centuries was a truly global project, orchestrated by a dedicated group of individuals as a complex expression of new colonial and scientific power that was making its way around the globe. For a handful of museums, the frenzy to establish bone collections became deeply influenced by significant global events—colonialism in Africa and Asia, U.S. expansion into the American West, and the First World War, happenings not always connected with the growth of physical anthropology collections. Yet these and other influences were, in fact, tangibly relevant to the expanding museum cultures in an era of growing and solidifying cultural institutions.

Aleš Hrdlička personified the race to collect bones for emerging museums in the United States; his vision for collecting helped build precedent for using the comparative study of skeletons to create a better understanding of what makes us human. He aspired to build a collection that would be the envy of his rivals in museums around the world. But those who care for his massive collections are bound to wrestle with his complex legacy for many years to come, as we continue to reassess the meaning of these remains for society today. His views on race continue to be criticized, and for many people, he represents the worst legacies of colonial anthropology in the nineteenth and early twentieth centuries. Indeed, while his project set off with the intention of showcasing racial difference, the collections he established, and the practice of studying them rigorously, helped establish a path for new scientific endeavors that ultimately undermined his own theories of racial classification. The project of understanding human prehistory and the origins of modern humans, on the other hand, continues to assume an increasingly prominent place in biological anthropology as a field. In recent years, new permanent exhibitions on human evolution opened at major museums in Chicago, New York, and Washington, DC.

New ideas, ethics, and science have also spurred healthy debate about medical museums in the United States, with the Mütter Museum emerging as the most prominent in the second half of the twentieth century. Indeed, even in more recent times, the Mütter Museum thrived as many other medical museums shuttered. The museum strives to create displays that mirror the exhibits of the institution's nineteenth-century origins. Yet, representing a trend in medical history museums, more recent advances in medicine and a harsh critique of collecting medical specimens compel curators to engage with visitors in ways that are foreign to the assumptions that buttressed the collecting of human skeletons in earlier generations. People continue to be attracted to the museum for different and challenging-to-define reasons. No matter the exact rationale or emotion, exhibits featuring human remains continue to maintain a strange and almost overpowering magnetism.

The collections described in this book, unlike in earlier eras, are rarely exhibited in public galleries. When included as part of new exhibits, however, human remains still draw crowds. Closing in 2014, the Smithsonian Institution's popular exhibit "Written in Bone: Forensic Files of the 17th-Century Chesapeake" hosted nearly nine million visitors. At least 680,000 of these visitors were students. Originally scheduled to be on view for only two years, so successful was the exhibition that the run extended to more than five years.[8]

Ultimately, the act of speaking for the dead proves a delicate, complex, sometimes tragic, and even wretched practice marred by ill or misguided intentions. Nevertheless, it is also a practice with clear potential for illumination about humanity, though stained by roots embedded in a history that causes much hurt for many ancestors of those who are described in this book. We must constantly reconsider the ethical implications of these collections in light of a troubled history and shifting intellectual paradigms, carefully weighing the benefits of these studies against the sensibilities of ancestral communities.

The collectors described in this book also characterized the bones they sent to museums in particular ways, with their written descriptions intimately tied to personal and culturally influenced ideas about race and the antiquity of humankind as it connected to old skeletons and, increasingly, fossils discovered overseas. While the scientists in this story viewed race as an undeniable reality, their shifting views on the shape of this reality reveal it to be an elastic construction heavily influenced by the changing contexts of each era.

The close study of these remains, colored as they were by racial science or nationalist claims, actually worked to disprove many theories espoused by early skull collectors. Perceived scientific realities connecting the size and shape of the skeleton to behavioral or intellectual attributes proved untenable. The science almost consumed itself. The rising tide of eugenics and scientific racism coincided with an extremely active period in collecting human remains for museums in the United States, shaping their storage and display in profound ways. While ideas changed over time, even shifting concepts behind the meaning and value of these collections left complex legacies. Eugenic ideas, in particular, seemed to echo and persist among bone collectors well after the field began falling out of favor in mainstream science.

While ethnography, archaeology, and linguistics largely moved away from museums as institutions, research stemming from studies on human remains collections in museums continued to be an active presence in science into the mid-twentieth century. As museums have transformed from collections- and research-oriented institutions to places that are more firmly centered on education and outreach, human skeletal collections confound museum leaders as they work to strike a balance between the desire to work with bones scientifically and the legitimate religious demand to see certain remains repatriated and reburied.[9]

During the late nineteenth and early twentieth centuries, museum leaders often bristled at proposals from curators to feature bodies in exhibits. Audiences would be appalled, they frequently argued, despite the fact that the reactions of museum visitors and newspaper editorialists routinely suggesting otherwise. Bones and mummified remains appealed to human curiosity on some base level, and the chance to view unusual bodies on display opened up audiences to other ideas in the exhibit. Despite popular exhibits, the contents in bone rooms were largely confined to the behind-the-scenes spaces in museums. When specimens were occasionally brought out for exhibitions, however, spectators were mysteriously drawn to them in large numbers. Why exactly people were so attracted to these galleries proves difficult to pin down. Exhibit curators frequently touted their benefit to other professionals and scholars, but they simultaneously attracted broader audiences, especially for world's fairs and new, permanent museum galleries. The Army Medical Museum's success was, at one point, attributed to collective mourning following the Civil War. Growing interest in prehistory emerging by the 1930s might be ascribed to new discoveries taking place around the world as well as to changing national attitudes and understandings about evolution. These exhibits, at their core, addressed fundamental, human questions: Where did we come from? Why are our bodies the way they are? and Why do some people look different from others? But read from a historical perspective, we see deep flaws in the answers offered for these pervasive questions in past exhibits. Indeed, people and their bodies were exploited in efforts to answer these key questions for both science and public audiences. But these same fraught questions still linger. Similar exhibits continue in museums today, in part because science connected to studying the dead promises to offer tantalizing new clues about who we are and where we have been.

In 1999, a market research firm working with the National Museum of Health and Medicine (NMHM—the successor to the Army Medical Museum) studied visitors' reactions to displays of human remains through in-depth surveys and focus-group interviews. Rather than expressing shock or disgust, most visitors had a rudimentary understanding of why a museum might choose to collect and exhibit a "specimen" for educational purposes and, in some cases, for scientific research. Those who took the survey widely reported a stronger emotional and intellectual reaction to the natural or "real" specimens shown to them for the study, compared to photographs or plastic reproductions of bodies or body parts. Much like their earlier counterparts who visited the exhibits described in this book, participants in this study were drawn to the actual human remains because they allowed them to ponder the nature of medicine, the human body, and death in a unique manner. Important to shaping visitors' reactions were the labels used in conjunction with the specimen or object—a telling reminder of the significance of context in the construction of our collective understandings of the human body and the essence of humanity itself.[10]

Since the time of the NMHM study, exhibits using new methods for preservation of human bodies have worked to usher in a resurgence of bodies on display. In particular, the *plastination* method associated with the wildly popular series of traveling exhibitions *Body Worlds* renewed the practice of exhibiting and gazing at human remains, bringing along with it new ethical questions related to the display of the human body for popular audiences. Emerging in the *Body Worlds*' sizable wake have been several poorly constructed and ethically dubious copycat exhibitions appearing on display in Las Vegas and Atlantic City casinos and in civic centers, rather than in museum galleries.

Even when approached with the best intentions, these exhibitions that edge into commercial, for-profit endeavors raise moral questions concerning the ownership of human remains. Does this constitute a moral quandary that is distinct from that of similar exhibits constructed in the name of science? Do we dishonor the dead by profiting from their remains? How best do we shape the fate of human remains in *any* effort to preserve and exhibit them? Who should make these important decisions, the scientists and curators who seek to exhibit these bodies or the living relatives of these past people? Ongoing debates on these subjects reveal that the issues are far from resolved.

REMEMBRANCE

What we learn by studying the history of these collections is that their stories speak to highly contested and challenging efforts to understand humankind through the creation of scientific categories and taxonomies. We also learn that for museums in the United States, even the distant human past represented an opportunity to illuminate the most central of American problems—race. The remains of the dead were frequently transformed into scientific resources that were considered crucial for understanding the problems of race and history, pushing aside other meanings commonly associated with bodies—death, ancestry, and medicine among them. These collections of human remains also expose as much about the living as they do about the dead. We learn that our rough binaries of colonial and anticolonial or racist and antiracist biographies of anthropologists largely prove untenable on closer inspection. Indeed, many anthropologists who are not typically associated with this story were occasionally bone collectors or proved otherwise complicit with the project in a variety of ways. We therefore must seek to understand both the history of these individuals and the institutions with which they became associated in a more nuanced and complex manner.

Partially due to the many ethical problems associated with these collections, anyone studying them must tread carefully. Contemporary scholars must approach these remains with the deep sense of humility and the sincere respect of people who recognize our own temporality in the human story. These collections, like anything else in a museum, possess a historical context connected to the evolution of intellectual networks, museum display, and the predominant sociocultural views defining each particular era. We can work to address some of these structural problems by enlisting the participation of living ancestors in the process of determining the fate of these remains, but repatriation and reburial does not work to undo the history behind these collections.

In the end, we learn that while the ideas about these skeletons are constantly evolving, some meaning tied to these bones transcends their justified use as objects for scientific study. Their essence is a humanity that surpasses changing historical contexts. In order to dignify them, we must confront this humanity at every level.

Despite the problems of translating what these remains are saying exactly, it is clear that they are still speaking. They offer us important lessons about our own mortality and humanity, serving to humble our absolute confidence in science. The exact nature of the lessons to be drawn from bone rooms serves as an ongoing point of widespread contention. Recent progress notwithstanding, this represents a problem with no clear end in sight.

If these now hushed voices could speak directly to the living, perhaps they would simply be intent on reminding us of their existence. They might try to humanize themselves against their portrayal as numbered scientific specimens. Maybe they would push back against the misconception that recent legislation has resolved all questions surrounding their position. The story of their collection, study, and exhibition leaves us with a legacy we must continually confront as we reassess science and society. Museums can serve as key spaces to attempt to come to terms with the colonial legacy attached to archaeology and anthropology, through partially redressing past wrongs while continuing the search for new knowledge. We owe it to the dead to carry on the conversation, as difficult as it may be, in an effort to better understand their legacy and contribution to our living world.

NOTES

ACKNOWLEDGMENTS

INDEX

NOTES

PROLOGUE

1. *New Ulm Post*, May 20, 1864, New Ulm Historical Society, New Ulm, Minnesota. U.S. Army records corroborate much of the story appearing in the *New Ulm Post*. Scouting parties of volunteers were ordered to search the region, and a firefight broke out when three Native Americans fired on a pair of soldiers. The soldiers returned fire, with bullets striking the single young man in the head and neck. Stephanie A. Makseyn-Kelley and Erica Bubnaik Jones, "Inventory and Assessment of Human Remains from the Historic Period Potentially Affiliated with the Eastern Dakota in the National Museum of Natural History," memorandum, April 24, 1996, Repatriation Office, National Museum of Natural History, Smithsonian Institution, Washington, DC.
2. The acting assistant surgeon, Alfred Muller, details the scalping and removal of specific body parts, but it was not until 1996 that physical anthropologists at the Smithsonian determined the extent of the individual's injuries, including a broken hip bone, cracked ribs, and cut marks on his right radius (an arm bone). Makseyn-Kelley and Jones, "Inventory and Assessment."
3. Alfred Muller, Acting Assistant Surgeon, U.S. Army, to Surgeon General, U.S. Army, March 26, 1866, Army Medical Museum (AMM) Records,

National Anthropological Archives, Smithsonian Institution, Washington, DC (hereafter AMM Records).
4. *New Ulm Post*, May 20, 1864, New Ulm Historical Society, New Ulm, Minnesota.
5. Mary Lethert Wingerd, *North Country: The Making of Minnesota* (Minneapolis: University of Minnesota Press, 2010), 296–297.
6. Alfred Muller to Army Medical Museum, March 26, 1866, AMM #13, Reel 1, AMM Records.
7. Muller even collected cracked bones that appeared to be from a white settler who perished in the earlier conflict, sending them to the Army Medical Museum. *Catalogue of the Surgical Section of the United States Army Medical Museum* (Washington, DC: Government Printing Office, 1866), 223.
8. Alfred Muller to Surgeon General, March 26, 1866.
9. "The . . . Indian belonged to one of the hostile Sioux tribes and was killed by soldiers in May 1864, about 12 miles south of this fort. I am sorry to state that the body was badly mutilated by soldiers and citizens before I was able to secure it; one of the hands and the scalp having been cut off and carried away, the lower jaw having fractured, . . . besides other unnecessarily treatment of the corpse." Alfred Muller to Surgeon General, U.S. Army, May 9, 1866, AMM Records.
10. I understand this group as the primary *discursive community* in this study. These individuals left behind a rich library of archival documents—museum memoranda, correspondence, field notes, and voluminous publications. Later on, a select few also left critical oral histories—adding to what we know of the development of this community.
11. Recent additions to the literature on the history of collecting skeletons for the purposes of studying race include Ann Fabian, *The Skull Collectors: Race, Science, and America's Unburied Dead* (Chicago: University of Chicago Press, 2010); Tony Platt, *Grave Matters: Excavating California's Buried Past* (Berkeley, CA: Heyday, 2011); Ian Tattersall and Rob DeSalle, *Race? Debunking a Scientific Myth* (College Station: Texas A&M University Press, 2011); and C. Loring Brace, *"Race" Is a Four-Letter Word: The Genesis of the Concept* (Oxford: Oxford University Press, 2005). The anthropologist Simon Harrison has also recently explored the charged connections between military trophy taking, race, and science. Simon Harrison, *Hunting and the Enemy Body in Modern War* (New York: Berghahn Books, 2012). Major

books on the history of archaeology and archaeological theory relevant to this study include Bruce G. Trigger, *A History of Archaeological Thought*, 2nd ed. (Cambridge: Cambridge University Press, 2006); Thomas C. Patterson, *Toward a Social History of Archaeology in the United States* (Fort Worth, TX: Harcourt Brace, 1995); and Alice B. Kehoe, *The Land of Prehistory: A Critical History of American Archaeology* (New York: Routledge, 1998). A useful guide to accumulations of skeleton remains at museums, universities, and beyond can be found in Christine Quigley, *Skulls and Skeletons: Human Bone Collections and Accumulations* (Jefferson, NC: McFarland, 2001). An important reference work on the history of physical anthropology is Frank Spencer, *A History of Physical Anthropology: An Encyclopedia*, 2 vols. (New York: Garland, 1997). Spencer also contributed a valuable work in Frank Spencer, *A History of Physical Anthropology, 1930–1980* (New York: Academic Press, 1982).

12. The remains of this individual have been repatriated. David Hunt, Smithsonian Institution, personal communication, December 3, 2009.

13. Michael Rhode and James Connor, "A Repository for Bottled Monsters and Medical Curiosities: The Evolution of the Army Medical Museum," in *Defining Memory: Local Museums and the Construction of History in America's Changing Communities*, ed. Amy Levin (Walnut Creek, CA: AltaMira, 2006), 177. Comparative anatomical collections started arriving at the museum in 1867.

14. The U.S. Civil War's implications for racial science and skull collecting is also explored in a comparative context by the anthropologist Simon Harrison. Harrison, *Dark Trophies: Hunting and the Enemy Body in Modern War* (New York: Berghahn Books, 2012), 93–106.

15. Quoted in David Hurst Thomas, *Skull Wars: Kennewick Man, Archaeology, and the Battle for Native American Identity* (New York: Basic Books, 2000), 58.

16. The focus of this book is the history of human skeletal remains in the United States. Other works focus on this history in other contexts, mapping the influence of European and American collecting practices on other parts of the world. See, for example, Martin Legassick and Ciraj Rassool, *Skeletons in the Cupboard: South African Museums and the Trade in Human Remains, 1907–1917* (Cape Town: South African Museum, 2000). Other works examine the connections between collecting human remains, trauma, and war. See Harrison, *Dark Trophies*. Recent additions to the literature

situate the history of collecting against the rise of professionalized museums in Europe, which deeply influenced the growth of museums in the United States. See Tony Bennett, *The Birth of the Museum: History, Theory, Politics* (London: Routledge, 1995); and Susan M. Pearce, *On Collecting, An Investigation into Collecting in the European Tradition* (London: Routledge, 1995).

17. Steven Conn, *Museums and American Intellectual Life, 1876–1926* (Chicago: University of Chicago Press, 1998), 5.

18. Exhibiting ideas on racial classification and the evolution of mankind took many forms during this period and emerged in numerous contexts outside museums. This study focuses on both the practice of collecting human skeletal remains and its influence on the history of race and science in the United States. Other important works chronicle other types of performances and expressions of visual culture outside of museums and international exhibitions. Jane R. Goodall, *Performance and Evolution in the Age of Darwin: Out of the Natural Order* (London: Routledge, 2002); Barbara Larson and Fae Brauer, eds., *The Art of Evolution: Darwin, Darwinisms, and Visual Culture* (Hanover, NH: Dartmouth College Press, 2009).

19. P. M. Jones, diary, January 20, 1901, photocopy of original, Accession No. 24, Hearst Museum of Anthropology, University of California, Berkeley.

20. This notion can also be seen in the use of the Civil War dead in crafting particular political and nationalistic agendas. "Without agendas, without politics, the Dead became what their survivors chose to make them." Drew Gilpin Faust, *This Republic of Suffering: Death and the American Civil War* (New York: Knopf, 2008), 269.

21. See George Stocking, *Victorian Anthropology* (New York: Free Press, 1987); and George Stocking, *The Ethnographer's Magic and Other Essays in the History of Anthropology* (Madison: University of Wisconsin Press, 1992). Only a decade before Stocking published *Victorian Anthropology*, the historian Thomas Haskell published his landmark study examining the professionalization of the social science disciplines. Thomas L. Haskell, *The Emergence of Professional Social Science* (Urbana: University of Illinois Press, 1977).

22. Bruce Kuklick explores early collecting for museums in the United States from the Old World civilizations of the Middle East. Bruce Kuklick, *Puritans in Babylon* (Princeton, NJ: Princeton University Press, 1996). Steven Conn offers several works important to the history of museums and mu-

seum anthropology. In his first book, Conn traces the rise and fall of museums as central figures to the growth of the American intellectual tradition. Conn, *Museums and American Intellectual Life*. In Conn's more recent work, he argues that museum objects have lost their centrality in presenting knowledge to museums visitors over the course of the twentieth century. Steven Conn, *Do Museums Still Need Objects?* (Philadelphia: University of Pennsylvania Press, 2010). Ira Jacknis traces the development of collecting from cultures of the Pacific Northwest as an example of broader collection-building trends in U.S. museums. Ira Jacknis, *The Storage Box of Tradition: Kwakiutl Art, Anthropologists, and Museums, 1881–1981* (Washington, DC: Smithsonian Institution Press, 2002).

23. George Stocking, *Objects and Others: Essays on Museums and Material Culture* (Madison: University of Wisconsin Press, 1985), 8.

24. The anthropologist Kathleen Fine-Dare places the practice of acquiring human remains within the broader systematic collecting of Native American material now falling under the Native American Graves Protection and Repatriation Act. Kathleen S. Fine-Dare, *Grave Injustice: The American Indian Repatriation Movement and NAGPRA* (Lincoln: University of Nebraska Press, 2002), 30–40. Louis Menand examines the practice of collecting and measuring skulls in the context of the study of race and human evolution in the long nineteenth century. Louis Menand, *The Metaphysical Club: A Story of Ideas in America* (New York: Farrar, Straus and Giroux, 2001). To date, the archaeologist and museum curator David Hurst Thomas has written one of the most complete accounts of the collection and study of human remains in the United States. Thomas, *Skull Wars*.

25. Spencer produced several major works, including Frank Spencer, ed., *A History of American Physical Anthropology, 1930–1980* (New York: Academic Press, 1982).

26. For a brief summary of these theoretical issues, see Katherine Ott, "The Sum of Its Parts: An Introduction to Modern Histories of Prosthetics," in *Artificial Parts, Practical Lives: Modern Histories of Prosthetics*, ed. Katherine Ott, David Serlin, and Stephen Mihm (New York: NYU Press, 2002), 4–5.

27. Erin Hunter McLeary, "Science in a Bottle: The Medical Museum in North America, 1860–1940" (Ph.D. diss., University of Pennsylvania, 2001), 14.

28. A brief summary of the literature surrounding death and burial in the United States can be found in Michael Kammen, *Digging Up the Dead: A*

History of Notable American Reburials (Chicago: University of Chicago Press, 2010), 22.
29. Fabian, *Skull Collectors*.
30. This claim was also made by a physical anthropologist living through the transition and is a question with which I wrestle in the epilogue. T. D. Stewart, "The Growth of American Physical Anthropology between 1925 and 1975," *Anthropological Quarterly* 48, no. 3 (1975): 193–204.
31. The National Park Service maintains several major databases with NAGPRA inventory data. These inventories suggest the total number of human remains at museums in the United States. "National NAGPRA Online Databases," National Park Service, Department of Interior. http://www.nps.gov/nagpra/ (accessed September 27, 2015).
32. Megan J. Highet, "Body Snatching and Grave Robbing: Bodies for Science," *History and Anthropology* 16, no. 4 (2005): 434.
33. Ibid.
34. This fact is crucial to those who hope to secure remains and affiliated funerary objects for repatriation, as legal guidelines have heretofore required cultural affiliation to be established before repatriation of human remains and sacred objects can occur. See Chip Colwell-Chanthaphonh, "Remains Unknown: Repatriating Culturally Unaffiliated Human Remains," *Anthropology News* 51, no. 3 (2010): 4.
35. Certain museums benefited from location and particularly energetic curators. The Hearst Museum of Anthropology at the University of California, Berkeley, for instance, gradually collected thousands of skeletons from shell mounds and burials discovered around the state in the late nineteenth and early twentieth centuries. The Field Museum, the American Museum of Natural History, and the Peabody Museum of Ethnography and Archaeology at Harvard also received queries and offers of skeletons stemming from accidental discoveries in proximity to the museum. Just as often, however, building collections was as much about competing on a global scale with other museums.

1. COLLECTING BODIES FOR SCIENCE

1. This book focuses on the physical anthropology and medical communities as they worked to build collections of human remains in major museums in the United States. Numerous other works explore collecting, researching, and displaying the body as it relates to race and prehistory. The leading

work in this new scholarship is Ann Fabian, *The Skull Collectors: Race, Science, and America's Unburied Dead* (Chicago: University of Chicago Press, 2010).
2. By the 1870s, a reliable postal network had been established in the United States. The Gold Rush, the Civil War, railroads, and the rapid growth of urban centers helped solidify modern communication networks. The relative ease of shipping packages allowed human remains to move about within the United States. For more on the creation of a modern postal network in the United States, see David M. Henkin, *The Postal Age: The Emergence of Modern Communications in Nineteenth-Century America* (Chicago: University of Chicago Press, 2006).
3. "Alaskan Mummies: What Capt. Hennig Found in the Aleutian Islands, an Indian Tradition the Bodies of a Chief and His Family That Have Been Preserved a Century and a Half Their Transfer to San Francisco. The Tradition. The Truth of the Story," *New York Times*, January 18, 1875, 5.
4. One book highlighting this kind competition in the Pacific Northwest, specifically among the Kwakiutl, is Ira Jacknis, *The Storage Box of Tradition: Kwakiutl Art, Anthropologists, and Museums, 1881–1981* (Washington, DC: Smithsonian Institution Press, 2002).
5. "Alaskan Mummies."
6. I have written more on this subject in S. J. Redman, "'What Self Respecting Museum Is without One?': The Story of Collecting the Old World at the Science Museum of Minnesota 1914–1988," *Collections: A Journal for Museum and Archives Professionals* 1, no. 4 (2005): 309–327.
7. "American Antiquities," *New York Times*, April 15, 1874, 4.
8. The best summary of these events can be found in James Snead, *Ruins and Rivals: The Making of Southwest Archaeology* (Tucson: University of Arizona Press, 2001).
9. Ronald F. Lee's important history of the American Antiquities Act of 1906 begins by chronicling the significance of the year 1879 as a starting point for efforts in American archaeological preservation. I follow his formulation for the significance of the year 1879 for American archaeology and anthropology. See Ronald F. Lee, "The Origins of the Antiquities Act," in *The Antiquities Act: A Century of American Archaeology, Historic Preservation, and Nature Conservation*, ed. David Harmon, Francis P. McManamon, and Dwight T. Pitcaithley (Tucson: University of Arizona Press, 2006), 15–34.
10. Ibid., 17.
11. Ibid., 18.

12. Aleš Hrdlička argued that American physical anthropology began in 1830 in Philadelphia with Samuel George Morton. C. Loring Brace, "The Roots of the Race Concept in American Physical Anthropology," in *A History of American Physical Anthropology, 1930–1980*, ed. Frank Spencer (New York: Academic Press, 1982), 17.
13. The best summary of Samuel George Morton's collecting and research is Fabian, *Skull Collectors*. Fabian details how Morton influenced other scholars, like those at the AMM, who in turn began to build their own collections, which soon outpaced Morton's personal collection of skulls.
14. Fabian, *Skull Collectors*, 176–177.
15. Morton's contacts acquired skulls through various means, including grave robbing, in order to contribute to Morton's growing collections. According to at least one account, several researchers endured great risk to obtain skulls, a theme explored in more depth in later chapters. Also see Emily Renschler and Janet Mongre, "The Samuel George Morton Cranial Collection: Historical Significance and New Research," *Expedition* 50, no. 3 (2008): 31.
16. David Hurst Thomas, *Skull Wars: Kennewick Man, Archaeology, and the Battle for Native American Identity* (New York: Basic Books, 2000).
17. Morton convinced another important American scientist, Louis Agassiz, that polygenesis had occurred. Louis Menand, "Morton, Agassiz, and the Origins of Scientific Racism in the United States," *Journal of Blacks in Higher Education* 34 (2001–2002): 111. C. Loring Brace carefully parses out Morton's polygenist viewpoint vis-à-vis other French anthropologists in the era, arguing that their positions were more akin to Morton's than to other American theorists'. Brace, "Roots of the Race Concept," 18–19. In another work, Brace offers a summary of the founding of the American School of Anthropology, including offering context for the work of Morton and his contemporaries. C. Loring Brace, *"Race" Is a Four-Letter Word: The Genesis of the Concept* (Oxford: Oxford University Press, 2005), 76–92.
18. Brace, "Roots of the Race Concept," 20.
19. John S. Michael, "A New Look at Morton's Craniological Research," *Current Anthropology* 29, no. 2 (1988): 349–350.
20. Stephen J. Gould, *The Mismeasure of Man* (New York: Norton, 1996).
21. Michael, "New Look at Morton's Craniological Research," 353; Menand, "Morton, Agassiz, and the Origins of Scientific Racism," 110–113.
22. A more complete summary of race theory in the early United States can be found in Bruce Dain, *A Hideous Monster of the Mind: American Race Theory in the Early Republic* (Cambridge, MA: Harvard University Press, 2002).

23. Michael, "New Look at Morton's Craniological Research," 353.
24. Quoted in Menand, "Morton, Agassiz, and the Origins of Scientific Racism," 110.
25. Lee D. Baker, *From Savage to Negro: Anthropology and the Construction of Race, 1896–1954* (Berkeley: University of California Press, 1998), 14–15.
26. Ales Hrdlička postulated as much in 1918. Brace, "Roots of the Race Concept," 17.
27. Ibid., 18–20.
28. Numerous overviews of the history of the AMM are in existence, but one of the most informative accounts of William Hammond's role in creating the museum is in Morris Leikind, "Army Medical Museum and the Armed Forces Institute of Pathology in Historical Perspective," *Scientific Monthly*, August 1954, 71–78. A more recent account is Michael Rhode and James Connor, "A Repository for Bottled Monsters and Medical Curiosities: The Evolution of the Army Medical Museum," in *Defining Memory: Local Museums and the Construction of History in America's Changing Communities*, ed. Amy Levin (Walnut Creek, CA: AltaMira, 2006), 177. Comparative anatomical collections started arriving at the museum in 1867.
29. Darwin wrote about his observations of human remains collections on display in Paris in Charles Darwin, *The Decent of Man and Selection in Relation to Sex*, 2nd ed. (New York: D. Appleton, 1909), 169–172. Thomas Jefferson's widely cited efforts as amateur archaeologist resulted in his observation, description, and collection of skeletal remains from American Indian graves. Thomas Jefferson, *Notes on the State of Virginia* (New York: Penguin Books, 1999), 100–105.
30. *List of Skeletons and Crania in the Section of Comparative Anatomy of the United States Army Medical Museum for Use during the International Exposition of 1876 in Connection with the Representation of the Medical Department U.S. Army* (Washington, DC: Army Medical Museum, 1876). This small catalogue of the AMM's collection, printed in time for the 1876 world's fair, lists over 1,100 examples of animal crania and another 1,155 individual animal skeletons in the collection. The skeletons, collected from throughout North America, represented animals ranging from large mammals encountered in the American West (such as pronghorn and buffalo) to small birds and reptiles.
31. Quoted in Robert S. Henry, *The Armed Forces Institute of Pathology: Its First Century, 1862–1962* (Washington, DC: Office of the Surgeon General Department of the Army, 1964), 55. This quote is attributed to Lt. Col. J. J. Woodward.

32. Quoted in ibid., 56.
33. These explanatory notes are significant as they provide the archival records critical for determining the provenance of the remains brought to the museum during this period. Leikind, "Army Medical Museum and the Armed Forces Institute of Pathology," 72.
34. Quoted in Henry, *Armed Forces Institute of Pathology*, 59. Various sources detail the history of the AMM. See specifically the introduction to the AMM Records, National Anthropological Archives, Smithsonian Institution, Washington, DC (hereafter AMM Records).
35. In a 1954 article about the history of the AMM, Morris Leikind describes the early museum as a sort of graduate school for medical officers. Morris, "Army Medical Museum and the Armed Forces Institute of Pathology," 73.
36. W. H. Forwood to Surgeon General, January 20, 1867, AMM Records.
37. See, for instance, AMM #8–11, AMM Records.
38. For Little Big Horn remains, see AMM #2120, AMM Records. This cranium arrived at the AMM with two conflicting accounts of its provenance. One set of AMM records indicates that the skull was obtained by a bugler who was the first man killed at the Battle of Little Big Horn. Another letter indicates that the soldier actually died two days following the battle.
39. AMM #136, AMM Records, for instance, is the skull of a Mandan, acquired from a cemetery at Fort Berthold, North Dakota.
40. AMM #400 and AMM #404–410, AMM Records.
41. AMM #637, AMM Records, for instance, was the skull of a white man, aged about thirty, hanged at Fort Benton, Montana, by a group described as "vigilantes."
42. Arthur Finkinsen and Dresden Drusticklund to The President of the Smithsonian Institution, May 24, 1872, AMM #916–917, AMM Records.
43. Jas. P. Kimball to George Otis, n.d., AMM #638, AMM Records.
44. William Henry Corbusier, *Soldier, Surgeon, Scholar: The Memoirs of William Henry Corbusier, 1844–1930*, ed. Robert Wooster (Norman: University of Oklahoma Press, 2003), 91.
45. Several years later, Corbusier returned to the San Carlos Agency and met with some of the individuals who had been relocated under his supervision. He writes, "Very little had been done for them, and they were only a little better off than they were in 1875." Ibid., 92.
46. AMM #411, AMM Records.
47. See, for instance, AMM #1802 and 1830, AMM Records.

48. See AMM #1756–1758, AMM Records.
49. G. P. Hachenberg to Surgeon General, October 20, 1879, AMM Records.
50. AMM #636, Reel 3, AMM Records. The specimen is described as the cranium of "Running Bear," shot in the Dakota Territories.
51. See, for example, Surgeon General's Office, *Reports on the Extent and Nature of the Materials Available for the Preparation of a Medical and Surgical History of the Rebellion* (Philadelphia: J. R. Lippincott, 1866). See also Alexander George Otis, *A Report of Surgical Cases Treated in the Army of the United States from 1865 to 1871* (Washington, DC: Government Printing Office, 1871).
52. For more on William Henry Holmes, including details on his personality, see Neil M. Judd, *The Bureau of American Ethnology: A Partial History* (Norman: University of Oklahoma Press, 1967), 12, 15, 23–25, 34, 39, 69, 70, 74.
53. Varying sources provide differing data on the exact number of remains transferred from the AMM to the Smithsonian Institution, and I rely here on the Smithsonian's most current count of remains included in this collection (including remains that have been repatriated since 1990).
54. According to the Smithsonian, the exact percentage is 79.3 percent. The information in this paragraph was obtained from Dave Hunt, Smithsonian Institution, personal communication, December 3, 2009.
55. The Smithsonian reports that 33.3 percent have been identified as females and 33.6 percent have been identified as males. Dave Hunt, Smithsonian Institution, personal communication, December 3, 2009.
56. P. T. Barnum, "Barnum on the World's Fair," *Chicago Daily Tribune*, March 6, 1890, 9.
57. On the tradition of collecting and displaying mummies from Egypt throughout the nineteenth century in the United States, see S. J. Wolfe and Robert Singerman, *Mummies in Nineteenth Century America: Ancient Egyptians as Artifacts* (Jefferson, NC: McFarland, 2009).
58. The anthropologist Eric Wolf elaborates on this Western-oriented notion in his important 1982 book about European contact with non-European societies: Eric Wolf, *Europe and the People without History* (Berkeley: University of California Press, 1982). The influence of Native Americans on the development of historical consciousness can be found in Steven Conn, *History's Shadow: Native Americans and Historical Consciousness in the Nineteenth Century* (Chicago: University of Chicago Press, 2004). See also

Maureen Kronkle, *Writing Indian Nations: Native Intellectuals and the Politics of Historiography, 1827–1863* (Chapel Hill: University of North Carolina Press, 2004); and Kerwin Klein, "Native Americans and the Burden of History," *Modern Intellectual History* 2, no. 3 (2005): 409–417.

59. The National Anthropological Archives maintains newspaper clippings originally gathered by the Smithsonian Institution's Department of Anthropology. These clippings come from around the country but are primarily from newspapers in the American West. Western newspapers reported most heavily on the discoveries and detail examples of temporary displays of mummies in local cities and towns. Box 12, Folder: Cliff Dwellers—Clippings A, Records of the Department of Anthropology, United States National Museum / National Museum of Natural History, Division of Ethnology, Manuscript and Pamphlet File, National Anthropological Archives, Smithsonian Institution, Washington, DC (hereafter RDA).

60. "A Discovery of Indian Mummies," *Harper's Magazine* 6, no. 8 (1887): 562.

61. Ibid. The same article in *Harper's* continues by introducing the possibility that living tribes may have been responsible for the burials: "The features of all these bodies would seem to preclude the possibility that they are Aztecs or Toltects. The weight of opinion of San Francisco archaeologists inclines to the belief that they are either Moquis or Zunis, as it is known that both these people have indulged in cave burial." Ibid.

62. "The Cliff Dwellers! The Wetherill Collection on Exhibition at Denver," clipping from Grand Junction, Colorado, newspaper, March 21, 1891, Box 12, Folder: Cliff Dwellers—Clippings A, RDA.

63. W. H. Holmes to B. K. Wetherill, January 31, 1890, Box 18, Folder: Habitations, Cliff Dwellings, RDA.

64. B. K. Wetherill to W. H. Holmes, March 3, 1890, ibid.

65. See Scott N. Morris to Samuel P. Langley, Secretary of the Smithsonian Institution, January 21, 1890, ibid.; Jas. A. Jones to J. W. Powell, January 29, 1890, ibid.; Echi Kneezell to Prof. Langley, October 28, 1890, ibid.

66. The body apparently represented the remains of a mummified child, though the newspaper reporting the story proclaimed it as evidence that the "Cliff Men" were naturally small in stature and said that nearby scientists had proclaimed it to be an adult. This article was published sometime after 1904. "Cliff Men Evidently Small: Remarkable Find of Forest Service Man in Gila Country Sent to Smithsonian Institution," newspaper clipping, n.d., Box 12, Folder: Cliff Dwellers—Clippings A, RDA.

67. "Relics of an Extinct Race: Magnificent Collection of Mr. McLoyd's Now on Free Exhibition in This City," newspaper clipping, April 28, 1892, ibid.
68. Kathleen S. Fine-Dare, *Grave Injustice: The American Indian Repatriation Movement and NAGPRA* (Lincoln: University of Nebraska Press, 2002), 99–100.
69. A now sizable literature exists on the World Columbian Exposition. Recent works range from popular nonfiction accounts of the fair to more specific (and more academic) studies of various aspects of the fair. One study exploring the display of human remains at the fair is Julie K. Brown, *Health and Medicine on Display: International Expositions in the United States, 1876–1904* (Cambridge, MA: MIT Press, 2009), 42–87.
70. Henry W. Haynes, "American Archaeology during the Past Ten Years," *American Journal of Archaeology* 4, no. 1 (1900): 17–39.
71. In 1890, for instance, the *Chicago Daily* ran an article noting advances in geology and archaeology that pushed back not only the arrival of humans but also glacial epochs. The article specifically references remains: "The discoveries of prehistoric remains on the Pacific coast, and especially in British Columbia, finished completely the last chance at a reasonable contention by the adherents of the older view [of a young earth theory]." In other words, the World Columbian Exposition did not introduce many fairgoers to the idea that the earth was old; evidence indicated that humans had evolved over thousands—if not millions—of years, and discoveries of human remains in North America indicated ancient human occupation in the Americas. "Antiquity of the Race," *Chicago Daily*, August 2, 1890, 13.
72. On the display of Indians of Peru at the World Columbian Exposition, the *New York Times* reads, "He [the organizer of the exhibit] has also made arrangements to take to Chicago a band of the wildest and most barbarous Indians in Peru, and has secured for them a great quantity of the native costumes and dresses, which will make an attractive display." "Peru at the World's Fair," *New York Times*, March 11, 1892, 6.
73. A more recent scientific interpretation of anthropometric data gathered by Boas for the fair can be found in R. L. Jantz, D. R. Hunt, A. B. Falsetti, et al. "Variation among North Amerindians: Analysis of Boas's Anthropometric Data," in "Special Issue on the Biological Anthropology of New World Populations," *Human Biology* 64, no. 3 (1992): 435–461.
74. Franz Boas to John Wesley Powell, September 21, 1892, Records of the Bureau of American Ethnology, National Anthropological Archives,

Smithsonian Institution, Washington, DC (hereafter BAE). In this letter, Boas informs Powell, the head of the BAE, that he will no longer be working at Clark University due to ongoing conflicts between the faculty and administration. Boas continues, "I shall go to Chicago this winter in order to work up the exhibit from Canada and in Physical Anthropology for the World's Fair. This work will probably keep me occupied probably until next autumn." Box 264, Series 1, Correspondence, Letters Received, Franz Boas, 1889–1909, BAE.

75. Franz Boas, *The Central Eskimo* (1888; repr., Lincoln: University of Nebraska, 1964).
76. Robert Louis Sullivan, *The Anthropometry of the Siouan Tribes* (New York: American Museum of Natural History, 1920).
77. Ibid., 96.
78. Franz Boas to W. H. Holmes, December 1, 1904, Box 264, Series 1, Correspondence, Letters Received, Franz Boas, 1889–1909, BAE.
79. Franz Boas to John S. Billings, May 6, 1887, AMM #2443–2445, Reel 2, AMM Records.
80. Quoted in Thomas, *Skull Wars*, 59–60.
81. Bandelier wrote in the preface to his novel that the story was "the result of eight years spent in ethnological and archaeological study among the Pueblo Indians of New Mexico." He added, "I have hoped to make the 'Truth about the Pueblo Indians' more accessible and perhaps more acceptable to the public in general." Adolf F. Bandelier, *The Delight Makers*, 2nd ed. (New York: Dodd, Mead, 1918), v–vi. F. W. Hodge, of the Smithsonian Institution, authored the preface to the second edition and describes the popular reception of the book. F. W. Hodge, preface to the second edition, ibid., vii–viii
82. Lummis goes so far as to lament that it had grown more fashionable for scribes in the United States to write about Africa than about North America. This trend, he argued, was making the Bushmen better known to the average citizen of the United States than American Indians were. He hoped his work would capture the imagination of popular audiences and draw them to unknown corners of the continent. Charles Fletcher Lummis, *Some Strange Corners of Our Country: The Wonderland of the Southwest* (New York: De Vinne, 1892); Frederick H. Chapin, *The Land of the Cliff-Dwellers* (Boston: Appalachian Mountain Club, W. B. Clarke, 1892). A brief account of all of these works also appears in Lee, "Origins of the Antiquities Act," 22.

83. After a brief loan of the collection to the University of Pennsylvania, the collection was purchased for a total of $14,500—an immense sum for the late nineteenth century. C. D. Hazzard to William Pepper, January 30, 1896, Box 24, Folder: Hazzard-Hearst Collection Correspondence, Gratacos–Hazzard/Hearst Collection, University Museum Archives, University of Pennsylvania, Philadelphia (hereafter UMA).
84. For an overview on the purchase, see Pete Dobemeier, "Understanding the Wetherill and Wilmarth Collections at the Colorado Historical Society," memorandum June 14, 2002, 2. Archival records documenting the transaction are housed in the University Museum Archives, University of Pennsylvania: Box 24, Folder: Hazzard-Hearst Collection Correspondence, Gratacos–Hazzard-Hearst Collection, UMA. Of the human remains in the Hazzard collection, only three sets of mummified remains were submitted to the University of California. The remainder of the human remains continued to be housed at the University of Pennsylvania. It is interesting to consider that the remains eventually deposited at the University of California would have been excavated from their original location, moved to Denver, shipped to Chicago, placed on display, shipped to Philadelphia, catalogued, and then shipped to San Francisco (then the location of the University of California Museum of Anthropology). All of this would have occurred between the 1880s and 1901. See list of objects given to Berkeley, 1901, memorandum, Box 25, Folder: Hazzard-Hearst Collection, The Hazzard Collection of Cliff-Dweller Material, Hazzard-Hearst Collection, UMA.
85. Human remains account for eighty-six catalogue numbers in the Hazzard collection. The total number of objects in the collection varies widely throughout published and archival documents. It seems likely that Hazzard inflated the number of objects in the collection to make it seem more dramatic and valuable. Museum officials, on the other hand, probably provided a more accurate count of around one thousand total objects. Catalogue, "Group A. Consisting of Human Remains," Box 24, Folder: Hazzard Catalogue, McCloyd-Graham Collection, 1892, Gratacos–Hazzard-Hearst Collection, UMA.
86. Floyd W. Sharrock, "The Hazzard Collection," unpublished manuscript, Box 26, Folder: Hazzard-Hearst Collection, UMA.
87. Hazzard's associates solicited a buyer as the fair ended. Hazzard and his associates noted in their letter to the director of the University Museum,

University of Pennsylvania, that they planned to write to the Smithsonian Institution, the Field Museum, the Peabody Museum, and the American Museum of Natural History. Hazzard's associates likely named these rival institutions as a technique to drum up a sense of competition and raise interest in the collection. Hector Alliot to Stewart Culin, May 14, 1894, Box 24, Folder: Hazzard-Hearst Collection Correspondence, Gratacos–Hazzard/Hearst Collection, UMA.

88. "Statement in Regard to Mr. Hazard's [sic] Collection, Made by Mr. Cushing, the Ethnologist, at Washington," memorandum, Box 26, Folder: Hazzard-Hearst Collection, Heye Collection, UMA.
89. John Joseph Flinn, *Official Guide to the World's Columbian Exposition* (Chicago: The Columbian Guide Company, 1893), 58. Additional information about the creation and display of the Wetherill and Wilmarth Collections can be found in another Colorado Historical Society (now History Colorado) memorandum. Richard H. Wilshusen, "Archaeological Ceramics Collections at the Colorado Historical Society: Their Past Histories and Future Uses in Exhibits, Research, and Instruction," memorandum, February 2004, Colorado Historical Society. Bridget Ambler, curator of material culture at the Colorado Historical Society, was kind enough to share this document with me.
90. Flinn, *Official Guide*, 58.
91. J. W. Buel, *The Magic City: A Massive Portfolio of Original Photographic Views of the Great World's Fair and Its Treasures of Art, Including a Vivid Representation of the Famous Midway Plaisance* (St. Louis: Historical Publishing, 1894).
92. "With Western Mummies," newspaper clipping, *The Philadelphia Record*, November 10, 1895, News Clippings, 1889–1981, Reel 1, UMA. Newspapers across Philadelphia were in agreement on this point; see "Primitive Civilizations: The Great Exhibition at the University of Pennsylvania and Its Wonderful Light on the Past," newspaper clipping, *Philadelphia Times*, November 10, 1895, ibid. See also "The Cliff Dwellers," *Public Ledger*, November 8, 1895, ibid.
93. "A Show over 2000 Years Old," newspaper clipping, *Philadelphia Times*, November 9, 1895, ibid.
94. "Some part of the wonderful tale of aboriginal life told by the Hazzard collection of objects from the cliff-dwellers of Colorado, which had just been arranged for public inspection, was interpreted to a large and delighted audience, which assembled in the library building of the University of Penn-

sylvania, yesterday afternoon." Newspaper clipping, *Philadelphia Press*, November 9, 1895, ibid. Several months later, it was reported that the collection "has attracted much attention, and has been visited by residents of nearly every State in the Union." Newspaper clipping, *Philadelphia Public Ledger*, January 30, 1896, ibid.

95. "Cliff Dweller and His Relics," newspaper clipping, *Philadelphia Evening Telegraph*, November 8, 1895, ibid.
96. Ibid. See also "The Cliff Dwellers," *Public Ledger*, November 8, 1895, ibid.
97. This quote comes from St. Louis exhibitor W. Maurice Tobin, quoted in Nancy J. Parezo and Don D. Fowler, *Anthropology Goes to the Fair* (Lincoln: University of Nebraska Press, 2007), 246.
98. United States National Museum, annual report for the year ending June 30, 1898, 3–4.
99. Ibid., 7.
100. Ibid., 4–24.
101. Ibid., 4.
102. Robert Oppenheim, "Revisiting Hrdlička and Boas: Asymmetries of Race and Anti-Imperialism in Interwar Anthropology," *American Anthropologist* 112, no. 1 (2010): 92–94.
103. A new book on the development of anthropology at Harvard University offers many related insights by tracing the history of one (albeit very important) center for anthropological research and training. See David L. Browman and Stephen Williams, *Anthropology at Harvard: A Biographical History, 1790–1940* (Cambridge, MA: Peabody Museum Press, Harvard University, 2013). For a brief biographical sketch of Earnest Albert Hooton, see 335–336.
104. Malinda Maynor Lowery has documented some of the many differences between Hooton and Hrdlička regarding Native American identity. See Malinda Maynor Lowery, *Lumbree Indians in the Jim Crow South: Race, Identity, and the Making of a Nation* (Chapel Hill: University of North Carolina Press, 2010), 186–187.
105. Details on Hrdlička's influence on the American anthropological community and his contribution to the decline of scientific racism can be found in Elazar Barkan, *The Retreat of Scientific Racism: Changing Concepts of Race in Britain and the United States between the World Wars* (Cambridge: Cambridge University Press, 1992), 97–100.
106. This quote, as well as the use of the word "megalomaniac," can be found in a letter written by Robert Lowie in reference to Hrdlička's possible

candidacy on the Executive Board of the American Anthropological Association. Robert Lowie to Clark Wissler, November 8, 1920, Box 1, Folder: Lowie, Robert, Outgoing, 1916–1922, Papers of Robert Lowie, Manuscripts Collections, Bancroft Library, University of California, Berkeley.
107. For more on Paul Broca and his place in the professionalization of biological anthropology, see Brace, "*Race*," 144–158.
108. On Hrdlička's responses to French and German influences, see Frank Spencer, *A History of American Physical Anthropology, 1930–1980* (New York: Academic Press, 1982), 5–6, 14–17.
109. Brace, "Roots of the Race Concept," 16–17; Brace, "*Race*," 222–239.
110. Brace, "Roots of the Race Concept," 21.
111. Thomas, *Skull Wars*, 59.
112. Franz Boas, "Some Recent Criticisms of Physical Anthropology," *American Anthropologist* 1, no. 1 (1899): 98.
113. Ibid., 100.
114. Ibid., 106.
115. Frank Spencer, *Ecce Homo: An Annotated Bibliographic History of Physical Anthropology* (New York: Greenwood, 1986), 308.
116. Ales Hrdlička, "Arrangement and Preservation of Large Collection of Human Bones for Purpose of Investigation," *American Naturalist* 34, no. 397 (1900): 9–10.
117. Ibid., 10.
118. "New Relics and Fossils: Accessions to the Natural History Museum," *New York Tribune*, October 28, 1900, 1.
119. The success of living exhibits at the 1883 Amsterdam International Colonial Exposition and the 1889 Exposition Universelle in Paris set the stage for massive displays of indigenous people in the ensuing decades. Parezo and Fowler, *Anthropology Goes to the Fair*, 6.
120. Ibid., 7.
121. Ibid., 49.
122. The narrative of the acquisition of Eskimo remains by Hrdlička and Boas for the American Museum of Natural History is particularly heart wrenching. When confronted by both the media and descendants of the Eskimos collected, the museum covered up the truth of the acquisition and even orchestrated a sham burial of remains. Kenn Harper, *Give Me My Father's Body: The Life of Minik the New York Eskimo* (South Royalton, VT: Steerforth, 1986), 91–97.

123. Parezo and Fowler, *Anthropology Goes to the Fair*, 319–323.
124. I focus on where Boas and Hrdlička came to differ in terms of their collecting, researching, and interpreting human remains. Other scholars have ably described how the two men arrived at asymmetrical ideas about race. Oppenheim, *Revisiting Hrdlička and Boas*, 92–103.
125. Aleš Hrdlička to William Henry Holmes, May 29, 1903, Papers of Aleš Hrdlička. National Anthropological Archives, Smithsonian Institution, Washington, DC.
126. Ales Hrdlička, *Directions for Collecting Information and Specimens for Physical Anthropology* (Washington, DC: Government Printing Office, 1904).
127. The historian Helen MacDonald points, in particular, to Joseph Barnard Davis's 1853 article "Hints for Collecting and Preserving the Bones of Ancient Skulls," published in *Gentleman's Magazine* in Britain. Heather MacDonald, *Human Remains: Dissection and Its Histories* (New Haven, CT: Yale University Press, 2006), 102.
128. Hrdlička, *Directions for Collecting*, 8.

2. SALVAGING RACE AND REMAINS

1. Frances Densmore to Aleš Hrdlička, September 12, 1918, Papers of Aleš Hrdlička, National Anthropological Archives, Smithsonian Institution, Washington, DC (hereafter Hrdlička Papers).
2. Frances Densmore to Aleš Hrdlička, January 7, 1918, ibid. Given the dates of the other letters appearing to be related to this same incident, I suspect this letter is misdated "1918" instead of "1919."
3. Densmore notes in her typewritten and expanded diaries, written in 1944, "Sept. 7 Went from Red Lake to Bemidji (Thence to Bena, and out on the region about 10 miles where some old graves had been washed out by the lowering of Winneb. Dam. Gathered bones, bits of pottery, and secured 1 or 2 skulls)." Frances Densmore, "Chronology of the Study and Presentation of Indian Music from 1893 to 1944," manuscript, MS 4250, Box 1, Folders 1–6: Diaries to Letters Received, Papers of Frances Densmore, National Anthropological Archives, Smithsonian Institution, Washington, DC (hereafter Densmore Papers).
4. Frances Densmore to Aleš Hrdlička, October 14, 1918, Hrdlička Papers.
5. Aleš Hrdlička to Frances Densmore, October 18, 1918, ibid.
6. Aleš Hrdlička to Frances Densmore, January 14, 1919, ibid.

7. Densmore's diaries confirm that she was in northern Minnesota in September 1918, but they contain little additional information. Frances Densmore, "Diary of Frances Densmore," January–December 1918, Densmore Papers.
8. Aleš Hrdlička to W. H. Holmes, October 3, 1913, Hrdlička Papers.
9. W. H. Holmes to Aleš Hrdlička, April 30, 1903, ibid.
10. Aleš Hrdlička to W. H. Holmes, May 29, 1903, ibid.
11. Ales Hrdlička, *Directions for Collecting Information and Specimens for Physical Anthropology* (Washington, DC: Government Printing Office, 1904), 5.
12. Ibid., 6.
13. Ibid., 7.
14. "Uncle Sam's New Islands," *Chicago Daily Tribune*, January 7, 1900, 34.
15. Samuel A. Barrett, "Excavation of Indian Burial Grounds Near Winters on Putah Creek," memorandum, December 20 and 21, 1905, 4–5, photocopy of original, Accession File 194, Hearst Museum of Anthropology, Berkeley, CA.
16. A useful discussion of Hrdlička's views on statistical analysis can be found in Juliet Marie Burba, "Whence Came the American Indians? American Anthropologists and the Origins Question, 1880–1935" (Ph.D. diss., University of Minnesota, 2006), 103–109.
17. Hrdlička, *Directions for Collecting*, 8.
18. Aleš Hrdlička to William Henry Holmes, August 26, 1903, Hrdlička Papers. See also the final copy of the letter: Aleš Hrdlička to William Henry Holmes, August 27, 1903, ibid.
19. Memorandum, Physical Anthropology, n.d. (probably around July 1904), ibid. This memorandum appears to have been enclosed with a letter written by Otis T. Mason on July 9, 1904. In the introductory letter, Mason credits Aleš Hrdlička as the author, but no date or signature is found on the document itself.
20. Smithsonian Institution, annual report, 1907, 45.
21. Following the discovery of the remains, those that could be preserved and shipped to the Army Medical Museum were added to the collections in Washington, DC. Washington Matthews, J. L. Wortman, and John S. Billings, "The Human Bones of the Hemenway Collection in the United States Army Medical Museum at Washington," in *Memoirs of the National Academy of Sciences*, vol. 6 (Washington, DC: Government Printing Office, 1893), 141–142.

22. At the same moment that Washington Matthews was recounting his visits to monuments in the American West, the American Association for the Advancement of Science appointed a small committee, "to memorialize to Congress to take the necessary steps for the preservation of archaeologie [*sic*] monuments on the public's lands of the United States." W. H. Petter to Alice C. Fletcher, August 1887, Box 7, Folder: Preservation of Antiquities 1887–1907, Papers of Alice Fletcher and Francis La Flesche, National Anthropological Archives, Smithsonian Institution, Washington, DC (hereafter Fletcher/La Flesche Papers).
23. I thank Lars Krutak at the Smithsonian Institution for advancing my thinking about the role of certain popular authors in shaping American consciousness on archaeology and prehistory in the American Southwest.
24. "Relics Are Vanishing," newspaper clipping, *New York Herald*, December 20, 1896, News Clippings, 1889–1981, Reel 1, University Museum Archives, University of Pennsylvania, Philadelphia, PA (hereafter UMA).
25. Alice C. Fletcher to Franz Boas, February 15, 1904, Box 7, Folder: Preservation of Antiquities 1887–1907, Fletcher/La Flesche Papers.
26. For more on the role of the Wetherill brothers' discovery of Mesa Verde and its role in the creation of both federal and state legislation, see Kathleen S. Fine-Dare, *Grave Injustice: The American Indian Repatriation Movement and NAGPRA* (Lincoln: University of Nebraska Press, 2002), 99–100.
27. The historian Denise D. Meringolo also notes Mesa Verde's position as one of the first park sites protected under the new American Antiquities Act of 1906. See, for instance, Denise D. Meringolo, *Museums, Monuments, and National Parks: Toward a New Genealogy of Public History* (Amherst: University of Massachusetts Press, 2012), 48–49.
28. Hal K. Rothman, *Devil's Bargains: Tourism in the Twentieth-Century American West* (Lawrence: University Press of Kansas, 1998), 113–142.
29. Works of fiction, too, influenced the notion of a mystic and distant past in the American West. Perhaps the best-known example of this genre was *The Delight Makers* by Adolph F. A. Bandelier, originally published in 1883. The book was successful with both the public and the scientific community. For more on how these works of fiction and the career of Edgar Hewett fit in the growth of archaeological tourism, see Rothman, *Devil's Bargains*, 1998.
30. "Restoring the Mummies of the Cliff Dwellers," *Los Angeles Herald*, July 2, 1905, 1.

31. Ronald F. Lee, "The Origins of the Antiquities Act," in *The Antiquities Act: A Century of American Archaeology, Historic Preservation, and Nature Conservation*, ed. David Harmon, Francis P. McManamon, and Dwight T. Pitcaithley (Tucson: University of Arizona Press, 2006), 29.
32. "Bill Passed by the U.S. Senate on April 27, 1904," Box 7, Folder: Preservation of Antiquities 1887–1907, Fletcher/La Flesche Papers.
33. Lee, "Origins of the Antiquities Act," 29.
34. Smithsonian Institution, annual report, 1907, 48–49.
35. Raymond Harris Thompson, "Edgar Lee Hewett and the Politics of Archaeology," in *The Antiquities Act: A Century of American Archaeology, Historic Preservation, and Nature Conservation*, ed. David Harmon, Francis P. McManamon, and Dwight T. Pitcaithley (Tucson: University of Arizona Press, 2006), 38.
36. I do not intend to portray Hewett's efforts as singular. As the movement to protect American antiquities gained momentum, supporting his efforts was the American Association for the Advancement of Science, the Archaeological Institute of America, and the American Anthropological Association. Box 7, Folder: Preservation of Antiquities 1887–1907, Fletcher/La Flesche Papers.
37. One aspect of Hewett's urgent call to action was the destruction of burial mounds by vandals and collectors. Edgar Lee Hewett, "Historic and Prehistoric Ruins of the Southwest and Their Preservation," n.p., n.d., 5, copy deposited in Box 7, Folder: Preservation of Antiquities 1887–1907, Fletcher/La Flesche Papers.
38. Hewett was widely known as a talented teacher, and at least one scholar attests that his pedagogical skills made him unique among archaeologists of the American Southwest. These skills, combined with a talent for political lobbying and his background in archaeology, put him in an ideal position to lobby for the preservation of American antiquities. James Snead, *Ruins and Rivals: The Making of Southwest Archaeology* (Tucson: University of Arizona Press, 2001), 77.
39. Lee, "Origins of the Antiquities Act," 31. See also Thompson, "Edgar Lee Hewett and the Politics of Archaeology," 35–47.
40. *Congressional Record*, 59th Cong., 1st sess. (May 17–June 8, 1906), vol. 40, part 8, 7331.
41. "Appendix: Essential Facts and Figures on the National Monuments," in Harmon, McManamon, and Pitcaithley, *Antiquities Act*, 288.

42. Smithsonian Institution, annual report, 1907, 53.
43. This particular pamphlet appears to have been sent to William H. Holmes at the Smithsonian and was preserved in the Smithsonian Institution's Department of Anthropology Records. The pamphlet describes finds at or around the site during the 1916 field season. A number of Beam's original photographs are preserved at the Denver Public Library. See also United States Railroad Administration, National Park Service, "Mesa Verde National Park Colorado," pamphlet, n.d., Box 18, Folder: Habitations: Cliff Dwellings, Records of the Department of Anthropology, United States National Museum / National Museum of Natural History, Division of Ethnology, Manuscript and Pamphlet File, National Anthropological Archives, Smithsonian Institution, Washington, DC.
44. Emily Renschler and Janet Mongre, "The Samuel George Morton Cranial Collection: Historical Significance and New Research," *Expedition* 50, no. 3 (2008): 32.
45. Aleš Hrdlička to Edgar Hewett, July 30, 1914, Hrdlička Papers.
46. Aleš Hrdlička to Edgar Hewett, August 4, 1914, ibid.
47. Aleš Hrdlička to W. H. Holmes, January 9, 1917, ibid.
48. Smithsonian Institution, annual report, 1918, 26.
49. Aleš Hrdlička to William H. Holmes, September 22, 1917, Hrdlička Papers.
50. Ibid.
51. Aleš Hrdlička to William H. Homes, April 9, 1918, ibid.
52. Alfred Kroeber's widely used textbook included sections on fossil humans, prehistory, and race. Both Kroeber and his mentor, Boas, largely encouraged their students to begin their careers in anthropology with a broad overview of the field—including a careful study of fossil humans, prehistory, and the existing racial classification theories of the era. A. L. Kroeber, *Anthropology: Race, Language, Culture, Psychology, Prehistory* (New York: Harcourt, Brace, 1948).
53. The literature surrounding Ishi is vast, and the meaning of his narrative has been hotly disputed over the course of the past decade. The classical account, solidifying Ishi's place in both the history of anthropology and the history of California is Theodora Kroeber, *Ishi in Two Worlds: A Biography of the Last Wild Indian in North America* (Berkeley: University of California Press, 1961). Theodora Kroeber, Alfred Kroeber's second wife, wrote about Ishi on numerous occasions. She was joined in editing a volume about his life by the archaeologist Robert Heizer. Robert F. Heizer and

Theodora Kroeber eds., *Ishi: The Last Yahi, a Documentary History* (Berkeley: University of California Press, 1979). The matter was revisited by Alfred Kroeber's sons in Karl Kroeber and Clifton Kroeber, eds., *Ishi in Three Centuries* (Lincoln: University of Nebraska Press, 2003). See also, most recently, Douglas Cazaux Sackman, *Wild Men: Ishi and Kroeber in the Wilderness of Modern America* (Oxford: Oxford University Press, 2010).

54. The story of Ishi's death, his autopsy, the whereabouts of his brain, and the subsequent process of its repatriation is chronicled in Orin Starn, *Ishi's Brain: In Search of America's Last "Wild" Indian* (New York: Norton, 2004), 28. Starn's book was especially important in reopening a wave of new investigation into Ishi's remains. Kroeber's quote was originally written in a letter on March 24, 1916. The same quote is cited in Sackman, *Wild Men*, 279.

55. Nancy Scheper-Hughes, "Ishi's Brain, Ishi's Ashes: Anthropology and Genocide," *Anthropology Today* 17, no. 1 (2001): 16.

56. This narrative is also discussed in Sackman, *Wild Men*, 70–75.

57. Several other Eskimos from Smith Sound had been brought to New York City in 1896, and nearly all of them had died of tuberculosis by 1901. Like Kishu, their bodies were then autopsied, and particular attention had been given to the study of their brains. See Aleš Hrdlička, "An Eskimo Brain," *American Anthropologist* 3, no. 3 (1901): 454.

58. Ibid., 486.

59. Ibid., 456.

60. Ibid., 484.

61. Ibid., 488.

62. This argument is echoed in Sackman, *Wild Men*, 263.

63. Alfred Kroeber to Aleš Hrdlička, October 27, 1916, National Museum of Natural History, Department of Anthropology, Accession Number 60884, Smithsonian Institution.

64. A brief recounting of this narrative is told in David Hurst Thomas, *Skull Wars: Kennewick Man, Archaeology, and the Battle for Native American Identity* (New York: Basic Books, 2000), 220–221. Thomas postulates that Kroeber may have sent Ishi's brain to the Smithsonian in order to cement his relationship with Hrdlička. Given the long-standing professional relationship between the two men, however, it is equally likely that Kroeber simply wanted to be rid of an object of little scientific use to his institution. Alternatively, Kroeber may have been emotionally bothered by the permanent reminder that he failed to keep an unspoken promise to his old friend

Ishi, who, given his culture's attitudes toward death, would clearly not have wanted an autopsy. It might be added that the Smithsonian's collection of human brains again increased in size just a few years later, when, in 1921, the museum received a private collection from a medical doctor in Washington, DC. Smithsonian Institution, annual report, 1921, 30–31.

65. Aleš Hrdlička to W. H. Holmes, December 20, 1916, Hrdlička Papers.
66. Aleš Hrdlička to Alfred Kroeber, December 20, 1916, ibid.
67. Alfred Kroeber to R. Rathbun, Assistant Secretary, Smithsonian Institution, January 5, 1917, ibid.
68. For more on Kroeber as a museum professional, see Ira Jacknis, "Alfred Kroeber as Museum Anthropologist," *Museum Anthropology* 17, no. 2 (1993): 27–32.
69. Kroeber to Rathbun, January 5, 1917.
70. Aleš Hrdlička to William H. Holmes, November 1, 1912, Hrdlička Papers.
71. Smithsonian Institution, annual report, 1912, 12.
72. Aleš Hrdlička to William H. Holmes, June 17, 1914, Hrdlička Papers.
73. T. Dale Stewart, interviewed by Pamela M. Henson, January–May 1975, 33–34, Oral History Project Interviews, Archives and Special Collections of the Smithsonian Institution, Washington, DC.
74. Ibid., 34.
75. Aleš Hrdlička to William H. Holmes, October 11, 1916, Hrdlička Papers.
76. In one annual report for the Smithsonian, for instance, the museum notes that Hrdlička was studying both the origins of North American Indians and the antiquity of humans in Europe. Smithsonian Institution, annual report, 1912, 11–12.
77. Smithsonian Institution, annual report, 1913, 28.
78. In 1910, for example, when pondering the value of a collection of available Peruvian skulls, Hrdlička explained to W. H. Holmes that the specimens "present many grades of artificial deformation which makes them specially desirable for the purposes of exhibition and comparison." Aleš Hrdlička to W. H. Holmes, December 8, 1910, Hrdlička Papers.
79. Aleš Hrdlička to W. H. Holmes. September 23, 1911, ibid.
80. Alfred Kroeber to Aleš Hrdlička, March 1, 1918, ibid.
81. Details about the missing skeletal material, the purchase of remains in Peru, and Hrdlička's claims about his work with Peruvian officials can be found in a small set of correspondence. See Otto Holstein to Aleš Hrdlička, March 15, 1913; Aleš Hrdlička to Otto Holstein, May 9, 1913; Aleš Hrdlička

to H. Clay Howard, United States Minister, Lima, Peru, May 9, 1913; and Aleš Hrdlička to Julio C. Tello, May 9, 1913; all in Box 106, Folder: Correspondence, Hrdlička Papers.
82. Aleš Hrdlička to Edgar L. Hewett, May 21, 1913, ibid.
83. Aleš Hrdlička to T. D. Stewart, March 29, 1941, ibid.
84. Aleš Hrdlička to W. H. Holmes, November 21, 1919, ibid.
85. Basic biographical information about Densmore can be found in Charlotte J. Frisbie, "Frances Theresa Densmore," in *Women Anthropologists: Selected Biographies*, ed. Ute Gacs, Aisha Khan, Jerrie McIntyre, and Ruth Weinberg (Westport, CT: Greenwood, 1988), 51–58.
86. The division between Alice Fletcher and the Boasian anthropologists is drawn by David Hurst Thomas in particular. While the division he draws is certainly accurate in many ways, the conceptualization of some scholars as outside the colonialist use of human remains collecting has proven to be misleading, if not inaccurate. See Thomas, *Skull Wars*.
87. Alice Fletcher to F. W. Putnam, March 20, 1884, Box 5: "1884 C–F" letters, UAV 677.38, Peabody Museum General Correspondence, Harvard University Archives, Cambridge, MA. I wish to thank Joanna Scherer, curator emeritus, Department of Anthropology, National Museum of Natural History, Smithsonian Institution, for directing me to this particular letter. Scherer is an indefatigable scholar of Fletcher, and she was kind enough to pass along this citation.
88. Ibid.
89. George G. Heye to Aleš Hrdlička, July 29, 1914, Hrdlička Papers.
90. Aleš Hrdlička to George G. Heye, July 30, 1914, ibid.
91. Hrdlička noted that the specimens possessed several interesting traits. In Hrdlička's estimation, they possessed intentional deformation of the head. Also, some of the females belonged to differing tribes. Finally, one of the specimens appeared to be a large, white male. Aleš Hrdlička to George G. Heye, March 2, 1915, ibid.
92. Alfred Kroeber to John P. Harrington, November 7, 1923, Reel 6, Correspondence, Papers of John P. Harrington, National Anthropological Archives, Smithsonian Institution, Washington, DC (hereafter Harrington Papers).
93. George G. Heye to Aleš Hrdlička, July 31, 1914, Hrdlička Papers.
94. Aleš Hrdlička to W. H. Holmes, December 10, 1910, ibid.
95. Aleš Hrdlička to W. H. Holmes, November 1, 1912, ibid.

96. Aleš Hrdlička to W. H. Holmes, May 26, 1920, ibid.
97. Henry Field to William Duncan Strong, December 12, 1927, Box 6, Folder: FER-FL, Papers of William Duncan Strong, National Anthropological Archives, Smithsonian Institution, Washington, DC.
98. *Eugenics, hygiene,* and *demography* were all complex and constantly shifting terms over the course of the first half of the twentieth century. For scholars of anthropology, these terms applied to the study of heredity, and some of them leveraged a growing understanding of biology to attempt to shape human populations. On the surface, these ideas could be interpreted as somewhat benign. A key group of extreme thinkers, however, pushed the eugenics movement toward a more sinister belief in an ability to shape populations on the basis of perceived notions of inherent advantages or disadvantages due to inherited characteristics. This included the belief that certain minority populations, or races, might be slowly bred out of larger populations.
99. Aleš Hrdlička to W. H. Holmes, March 21, 1912, Hrdlička Papers.
100. Smithsonian Institution, annual report, 1913, 23.
101. Leon F. Whitney to Aleš Hrdlička, March 31, 1926, Hrdlička Papers.
102. Aleš Hrdlička to Leon F. Whitney, April 3, 1926, ibid.
103. Lee D. Baker, *From Savage to Negro: Anthropology and the Construction of Race, 1896–1954* (Berkeley: University of California Press, 1998), 93.
104. The papers read before the congress were compiled in two published volumes. *Eugenics, Genetics and the Family: Scientific Papers of the Second International Congress of Eugenics,* vol. 1 (Baltimore: Williams and Wilkins, 1923); and *Eugenics in Race and State: Scientific Papers of the Second International Congress of Eugenics,* vol. 2 (Baltimore: Williams and Wilkins, 1923).
105. *Eugenics, Genetics and the Family,* 1.
106. Ibid., 2.
107. Ales Hrdlička, *The Old Americans* (Baltimore: William and Wilkins, 1925), 1.
108. Ibid., 3.
109. Franz Boas, *Changes in Bodily Form of Descendants of Immigrants* (New York: Columbia University Press, 1912).
110. Hrdlička labored on the book for years, writing to Arthur Keith, "I really gave 'The Old Americans' the best that was in me." Aleš Hrdlička to Arthur Keith, January 9, 1926, Hrdlička Papers.
111. A description of the exhibition can be found in an undated memo, probably created around the closing of the exhibit. Undated memo, Correspondence,

Folder: International Congress of Eugenics, ibid. This memorandum was presumably written by Hrdlička to the organizers of the exhibition in 1921.

112. "Preparation of Exhibits Illustrating the Natural History of Man," in Smithsonian Miscellaneous Collections, Vol. 65. Washington, DC, Smithsonian Institution, 1916, 59.

113. Undated memo, Correspondence, Folder: International Congress of Eugenics, Hrdlička Papers.

114. Ibid.

115. Aleš Hrdlička to Leon F. Whitney, n.d., Hrdlička Papers. The letter is undated but was probably written in September 1927.

116. Report of the President of the American Eugenics Society, Inc., June 26, 1926, ibid.

117. In speaking in particular about collections from the American Southwest, George H. Pepper, the director of the University Museum, University of Pennsylvania, noted when the Hazzard collections were acquired for his museum following the Columbian World Exposition, these kinds of collections were unique. By 1908, he noted that collections of antiquities from the American Southwest were commonplace in museums across the United States. He wrote, "I think that the price paid was about $4,000 [actually $14,500]; if this was the figure it certainly was all that the collection is worth. At that time little of this class of material represented in Museum collections but now almost every Museum had a good showing." George H. Pepper to George B. Gordon, October 7, 1908, Box 24, Folder: Hazzard-Hearst Collection Correspondence, Gratacos Hazzard/Hearst Collection, UMA

118. F. W. Hodge to John P. Harrington, March 28, 1924, Reel 6, Correspondence, Harrington Papers.

3. THE MEDICAL BODY ON DISPLAY

1. Many of these museums have since disappeared, but several continue to operate. In the United States, these museums include the Mütter Museum, the Army Medical Museum (now the National Museum of Health and Medicine), the Warren Anatomical Museum of the Francis A. Countway Library of Medicine at Harvard Medical School (founded 1847), and the Indiana Medical History Museum (founded 1895). In Canada, a medical museum associated with the Montreal General Hospital grew into a full-fledged collection by the late nineteenth century. These museums, especially the Mütter Museum—a part of the College of Physicians of Philadelphia—

mimicked the much older anatomical displays at the Royal College of Surgeons of England. The International Association of Medical Museums emerged in 1906.

2. For records on the acquisitions made by the Mütter Museum, see College of Physicians of Philadelphia, Office of the Curator, "Museum Catalogue," [1849–1851?], Accession 1991-106, Mütter Museum, College of Physicians of Philadelphia, Philadelphia, PA (hereafter MM); and "Catalogue of the Mütter Museum of the College of Physicians of Philadelphia," [1884–1941?], 3 vols., Accession 1991-119, ibid.

3. Obvious disadvantages have long been inherent to either drying or preserving a specimen in liquid. Drying most soft tissue changed the shape and texture. Most liquids that were used to preserve soft tissue, while maintaining the overall size and shape of most specimens, changed the color of the sample. In light of these obvious problems, it is easy to see why collectors were eager to gather skeletal materials that were easier to preserve and, once dried and cleaned, typically maintained their shape, texture, and size over long periods. In comparison to the clear defects of either drying or fluid preservation, models possessed an obvious appeal.

4. Erin Hunter McLeary, "Science in a Bottle: The Medical Museum in North America, 1860–1940" (Ph.D. diss., University of Pennsylvania, 2001), 5.

5. *Bulletin of the International Association of Medical Museums* 1–8 (1907–1920): 1.

6. As noted in the body of this chapter, the Army Medical Museum opened its doors to the public much earlier than the Mütter Museum did.

7. For more on space, language, death, and the origins of the medical gaze, see Michel Foucault, *The Birth of the Clinic: An Archaeology of Medical Perception* (New York: Pantheon Books, 1973).

8. For additional context on these radical transformations in American medicine, see Lester S. King, *Transformations in American Medicine from Benjamin Rush to William Osler* (Baltimore: Johns Hopkins University Press, 1991).

9. Historians of science, medicine, and anatomy have reported that a comparatively small archive exists for either the Mütter Museum or the College of Physicians. The historian Whitfield J. Bell, Jr., lamented in 1987, "Of personal correspondence regrettably little seems to have survived," noting, "but, then, as the College was a local institution and the Fellows regularly saw one another on Spruce and Pine Streets, it is unlikely that they would have discussed College business in private letters." In reconstructing the

history of researching and displaying the medical body at this particular institution, I rely heavily on limited curatorial records, newspaper reports, and published records of the college. See Whitfield J. Bell, Jr., *The College of Physicians of Philadelphia: A Bicentennial History* (Canton, MA: Science History, 1987), viii. A recent overview of medical museums in the United States and Europe is in Samuel J. M. M. Alberti and Elizabeth Hallam, eds., *Medical Museums: Past, Present, Future* (London: Royal College of Surgeons of England, 2013).

10. Bell, *College of Physicians of Philadelphia*, 115.
11. Several valuable accounts of the formation of the Mütter Museum have appeared in print. Laura Lindgren, ed., *Mütter Museum, Historical Medical Photographs* (New York: Blast Books, 2007); Gretchen Worden, *The Mütter Museum of the College of Physicians of Philadelphia* (New York: Blast Books, 2002); Nancy Moses, *Lost in the Museum: Buried Treasures and the Stories They Tell* (Plymouth, UK: AltaMira, 2008), 61–64; *Transactions of the College of Physicians of Philadelphia* 9 (1887): clxxiv–clxxiv. Also see, most recently, Kathleen R. Sands and Elinor G. Hickey, *The College of Physicians of Philadelphia (Images of America)* (Charleston, SC: Arcadia, 2012).
12. Bell, *College of Physicians of Philadelphia*, 118.
13. Samuel J. M. M. Alberti and Elizabeth Hallam, introduction to *Medical Museums*, 3.
14. For the anatomy act, see William Smith Forbes, "History of the Anatomy Act of Pennsylvania" (1867), in *Body Snatching: The Robbing of Graves for the Education of Physicians in Early Nineteenth Century America*, by Suzanne M. Shultz (Jefferson, NC: McFarland, 2005), 111–116. The discourse around the use of living organisms for medical studies, known as the vivisection debates, was not nearly as heated in the United States as it was in Europe. Nevertheless, the debate surrounding the appropriate use of the body, both living and dead, for medical dissection was present in the United States, and it lingered for decades following the passage of anatomy acts. For an example, see George Hamilton, "Thoughts upon Vivisection, with Reference to Its Restriction by Legislative Action," *Transactions of the College of Physicians of Philadelphia* 5 (1881): 103–119.
15. Alfred Stillé, "Remarks Made by the President," *Transactions of the College of Physicians of Philadelphia* 6 (1883): xliii–xliv.
16. Author Laura Lindgren notes that the practice of displaying various medical conditions grew out of a long history of medical illustration. The prac-

tice of painting the human form eventually gave rise to the medical illustration, which soon influenced the practice of medical photography. Laura Lindgren, introduction to *Mütter Museum*, 14.
17. Quoted in Robert S. Henry, *The Armed Forces Institute of Pathology: Its First Century, 1862–1962* (Washington, DC: Office of the Surgeon General Department of the Army, 1964), 56.
18. Ibid., 28.
19. "Traced Arm Lost in War," *New York Times*, July 24, 1907, 1.
20. Quoted in Henry, *Armed Forces Institute of Pathology*, 56.
21. Michael Kammen, *Digging Up the Dead: A History of Notable American Reburials* (Chicago: University of Chicago Press, 2010), 114.
22. On S. Weir Mitchell, see William K. Beatty, "S. Weir Mitchell and the Ghosts," *Journal of the American Medical Association* 220, no. 1 (1972): 76–80; and D. J. Canale, "S. Weir Mitchell's Prose and Poetry on the American Civil War," *Journal of the History of the Neurosciences: Basic and Clinical Perspectives* 13, no. 1 (2004): 7–21. The most recent biography on Mitchell can be found in Nancy Cervetti, *S. Weir Mitchell, 1829–1914: Philadelphia's Literary Physician* (University Park: Pennsylvania State University Press, 2012).
23. Originally published in *Atlantic Monthly* in 1866, the text was republished in an expanded volume of fiction. S. Weir Mitchell, *The Autobiography of a Quack and the Case of George Dedlow* (New York: Century, 1900).
24. Michael Rhode, "An Army Museum or a National Collection?," in Alberti and Hallam, *Medical Museums*, 191–196.
25. Lindgren, introduction to *Mütter Museum*, 22.
26. Julie K. Brown, *Health and Medicine on Display: International Expositions in the United States, 1876–1904* (Cambridge, MA: MIT Press, 2009), 1–5.
27. Daniel T. Rodgers, *Atlantic Crossings: Social Politics in the Progressive Age* (Cambridge, MA: Harvard University Press, 2000).
28. An example of the kinds of public health campaigns, one that combined both major and minor exhibitions together with literature campaigns, is the sex-education efforts of the federal government in the United States. See Alexandra M. Lord, *Condom Nation: The U.S. Government's Sex Education Campaign from World War I to the Internet* (Baltimore: Johns Hopkins University Press, 2010).
29. Brown, *Health and Medicine on Display*, 25–26.
30. Ibid., 26.

31. Ibid., 26–27.
32. Julie Brown includes this information in a note (52). Brown, *Health and Medicine on Display*, 220.
33. Ibid., 60–64.
34. Ibid., 80–81. In the AMM's 1876 displays, the museum used the term "comparative anatomy" to mean comparisons with nonhuman animals. Army Medical Museum, *List of Skeletons and Crania in the Section of Comparative Anatomy of the United States Army Medical Museum for Use during the International Exhibition of 1876 in Connection with the Representation of the Medical Department U.S. Army* (Washington, DC: Army Medical Museum, 1876).
35. Aleš Hrdlička, *Tuberculosis among Certain Indian Tribes of the United States*, bulletin 42 (Washington, DC: Government Printing Office, Smithsonian Institution, Bureau of American Ethnology, 1909).
36. Brown, *Health and Medicine on Display*, 178–179.
37. Ibid., 180–181.
38. Bell, *College of Physicians of Philadelphia*, 133–135.
39. See *Transactions of the College of Physicians of Philadelphia* 1 (1875): 4–5.
40. Bell, *College of Physicians of Philadelphia*, 134. The autopsy is also the subject of volume 1 of the *Transactions of the College of Physicians of Philadelphia*.
41. "Chang and Eng," *New York Times*, February 11, 1874, 1.
42. *Transactions of the College*, 1:5.
43. "Records of the Autopsy of Chang and Eng Bunker," 1874–1875, Accession 1993-051, Committee on the Mütter Museum, MM.
44. In addition to the autopsy report itself, medical journals reported on the autopsy given to Chang and Eng, or the "monster," at the museum. See "The Siamese Twins," *British Medical Journal* 1, no. 689 (1874): 359–363.
45. "Chang and Eng," 1.
46. Limited catalogue information about Chang and Eng can be found in "Catalogue of the Mütter Museum of the College of Physicians of Philadelphia," [1884–1941?], vol. 2, Accession 1991-119, MM.
47. Bell, *College of Physicians of Philadelphia*, 136.
48. "Siamese Twins," 362.
49. Megan J. Highet, "Body Snatching and Grave Robbing: Bodies for Science," *History and Anthropology* 16, no. 4 (2005): 424–425.
50. The skeleton of the giant has been displayed next to the remains of a woman named Mary Ashberry, a who had achondroplastic dwarfism. Ashberry, who

stood three feet six inches tall, provided a striking contrast with the massive skeleton of the giant. Worden, *Mütter Museum*, 10, 183–184. See also Ella N. Wade, "A Curator's Story of the Mütter Museum and College Collections," *Transactions and Studies of the College of Physicians of Philadelphia* 42, no. 2 (1974): 126.

51. One archival document, in particular, provides a snapshot of medical museum collecting between about 1849 and 1851. College of Physicians of Philadelphia, Office of the Curator, "Museum Catalogue" [1849–1851?], Accession 1991-106, MM.

52. In 1911, for example, Joseph Tunis, a physician, described in detail his method for preserving a collection of sections of human heads. He had prepared the sections for the Mütter Museum. The sections were photographed, labeled, and utilized extensively in lantern slides for teaching demonstrations. Joseph P. Tunis, "Description of a Series of Frontal and Sagittal Sections of the Adult Human Head Recently Acquired by the Mütter Museum," *Transactions of the College of Physicians of Philadelphia*, 3rd ser., 33 (1911): 363–365.

53. Wade, "Curator's Story," 122–123.

54. The Mütter Museum, for example, maintained a long-standing relationship with the American Surgical Association (ASA). The ASA, over the course of the twentieth century, donated a large series of portrait photographs of its members and subsequently revisited these photographs, which the Mütter Museum preserved. Joseph McFarland to Walter Estell Lee, April 24, 1942, Folder: American Surgical Association, Curatorial Files, MM.

55. Hayes D. Agnew, "Annual Address of the President," *Transactions of the College of Physicians of Philadelphia* 12 (1890): xxviii.

56. "Catalogue of the Mütter Museum of the College of Physicians of Philadelphia," [1884–1941?], vol. 1, Accession 1991-119, MM.

57. Bell, *College of Physicians of Philadelphia*, 206.

58. Hayes D. Agnew, "Annual Address of the President," *Transactions of the College of Physicians of Philadelphia* 11 (1889): xxx.

59. "Appendix: Abstract for the Report of the Committee on the Mütter Museum, 1902," *Transactions of the College of Physicians of Philadelphia* 24 (1902): 321–322.

60. James Cornelius Wilson, "Annual Address of the President," *Transactions of the College of Physicians of Philadelphia* 38 (1916): 7.

61. "Appendix: Abstract of the Report of the Committee on the Mütter Museum, 1903," *Transactions of the College of Physicians of Philadelphia* 25 (1903): 179.
62. "Appendix: Abstract of the Report of the Committee on the Mütter Museum, 1904," *Transactions of the College of Physicians of Philadelphia* 26 (1904): 305.
63. "Appendix: Abstract of the Report of the Committee on the Mütter Museum, 1905," *Transactions of the College of Physicians of Philadelphia* 27 (1905): 231.
64. "There have also been numerous visits to the Museum by Fellows of the College, as well as by other members of the profession and laity residing in this Commonwealth and in various parts of this, and of foreign countries." "Report on the Committee on the Mütter Museum," *Transactions of the College of Physicians of Philadelphia* 36 (1914): 370.
65. "Report of the Committee on Mütter Museum for 1920," *Transactions of the College of Physicians of Philadelphia* 42 (1920): 452.
66. "Annual Reports of the Committees of the Mütter Museum and College Collections," *Transactions and Studies of the College of Physicians of Philadelphia* 12 (1945): 149.
67. Bell, *College of Physicians of Philadelphia*, 150.
68. Ales Hrdlička indicates this point when he argues that scholars concerning themselves with anthropometry should "avoid the inclusion of any individuals who may have been affected by some pathological condition sufficiently to suffer a material alteration in their measurements." Ales Hrdlička, *Anthropometry* (Philadelphia: Wistar Institute of Anatomy and Biology, 1920), 46.
69. For more on racism in medical training, dissection, and body collecting, see Robert L. Blakely and Judith M. Harrington, eds., *Bones in the Basement: Postmortem Racism in Nineteenth-Century Medical Training* (Washington, DC: Smithsonian Institution Press, 1997).
70. Quoted in Bell, *College of Physicians of Philadelphia*, 150–151.
71. Gretchen Worden, "The Hyrtl Skull Collection," *Transactions and Studies of the College of Physicians of Philadelphia* 17 (1995): 107.
72. Quoted in ibid., 102.
73. W. J. McGee to Guy Hinsdale, February 14, 1894, Folder: Muiz Collection of Trephined Skulls—Correspondence, Curatorial Files, MM. More information about the acquisition of the Muiz collection can be found in

"Catalogue of the Mütter Museum of the College of Physicians of Philadelphia," [1884–1941?], vol. 1, 467, Accession 1991-119, MM.

74. A photograph and brief description of the casts can be found in Worden, *The Mütter Museum*, 2002: 93.

75. Although the memorandum in the file is undated, clues in the text point to the acquisition of the collection of crania as taking place between 1892 and 1914. The text of the memorandum cites Rudolph Virchow's *Crania Ethnica America*, first published in 1892, and the donor of the collection, S. Weir Mitchell, died in 1914. "Indian Skulls from Burial Mounds in Illinois, near St. Louis, Missouri and Geneva Lake, Wisconsin. Sixty Indian Skulls Deposited by Dr. S. Weir Mitchell," memorandum, n.d., Accessions 1006.200 and 1006.147, Folder: Indian Skull Collection, Curatorial Files, MM.

76. Gretchen Worden, who was serving as the director of the Mütter Museum, estimated the date of this particular acquisition to be the late nineteenth century. This would be consistent with the date I have estimated for the related memorandum. Gretchen Worden to Dave Grignon, July 9, 1999, Folder: Indian Material, Curatorial Files, MM.

77. "Indian Skulls from Burial Mounds in Illinois."

78. Announcement for reception to the American Medical Association, June 5, 1897, Accession 1991-119-13, CPP 10/0007-03, Box 1, Letterbooks, 1858–1939, vol. 1, College of Physicians of Philadelphia–Committee on the Mütter Museum, MM.

79. *British Medical Journal* 2 (1900): 198.

80. "The College of Physicians of Philadelphia," *British Medical Journal* 2, no. 2554 (1909): 1703.

81. Ibid., 1703.

82. W. W. Keen, "The College of Physicians of Philadelphia: Its Library and Its New Building," *British Medical Journal* 2, no. 2546 (1909): 1161–1163.

83. Ibid., 1163.

84. "Appendix: Abstract of the Report of the Committee on the Mütter Museum, 1908," *Transactions of the College of Physicians of Philadelphia* (1908): 242.

85. Bell, *College of Physicians of Philadelphia*, 151.

86. Moses, *Lost in the Museum*, 59.

87. Competing sex-education campaigns in the United States, for example, generally relied on curriculum development, temporary displays, and mass mailings. Medical museums are conspicuously absent in postwar public health campaigns. See Lord, *Condom Nation*.

88. Lindgren, introduction to *Mütter Museum*, 14.
89. Ibid., 13.
90. Robert D. Hicks, "The Disturbingly Informative Mütter Museum," in Alberti and Hallam, *Medical Museums*, 172.
91. Lindgren, introduction to *Mütter Museum*, 22.

4. THE STORY OF MAN THROUGH THE AGES

1. Robert Rydell offers basic information about fairs as well as compelling analysis about their significance. In summarizing the fairs, I have relied on his chapter on the California fairs in his classic account of world's fairs in the United States. Robert Rydell, *All the World's a Fair: Visions of Empire at American International Expositions, 1876–1916* (Chicago: University of Chicago Press, 1987). A more recent and specific exploration can be found in Matthew F. Bokovoy, *The San Diego World's Fairs and Southwestern Memory, 1880–1940* (Albuquerque: University of New Mexico Press, 2005). Bokovoy discusses the physical anthropology displays in particular in his third chapter, 87–113. Bokovoy relies heavily on published primary and secondary accounts of the fair, including archival materials from the San Diego Historical Society. In this chapter, I rely on published primary and secondary accounts as well as the records of the Smithsonian Institution, which preserved the majority of the original archival material related to the development of the exhibition.
2. "These exhibits were in preparation for over three years. They are original and much more comprehensive than any previous exhibits in this line, either here or abroad." See "Preparation of Exhibits Illustrating the Natural History of Man," in *Smithsonian Miscellaneous Collections*, vol. 65 (Washington, DC: Smithsonian Institution, 1916), 55.
3. Ales Hrdlička, "The Division of Physical Anthropology at the Panama-California Expos, San Diego," unpublished manuscript, Correspondence 1912–1915, San Diego Exposition, Papers of Aleš Hrdlička, National Anthropological Archives, Smithsonian Institution, Washington, DC (hereafter Hrdlička Papers). This document, reprinted in numerous versions throughout archival materials, was eventually published as a catalogue and as the following article: Ales Hrdlička, "An Exhibit in Physical Anthropology," *Proceedings of the National Academy of Sciences of the United States of America* 1, no. 7 (1915): 407–410. The exhibit catalogue was published as

Ales Hrdlička, *A Descriptive Catalog of the Section of Physical Anthropology Panama-California Exposition 1915* (San Diego: National Views, 1915).
4. Rydell, *All the World's a Fair*, 219.
5. A recent comparison of the ideas found in the *Science of Man* building with those brought forward by Franz Boas, Alfred Kroeber, and others can be found in Douglas Cazauz Sackman, *Wild Men: Ishi and Kroeber in the Wilderness of Modern America* (Oxford: Oxford University Press, 2010), 262–265.
6. James Snead, *Ruins and Rivals: The Making of Southwest Archaeology* (Tucson: University of Arizona Press, 2001), 77.
7. Hewett had worked with Holmes and Hrdlička while working at the Bureau of American Ethnology in an unsalaried position while he lobbied for the passage of the American Antiquities Act. Ibid., 77.
8. C. D. Walcott to Aleš Hrdlička, March 30, 1912, Box 22, Folder 5: Expositions: Panama California Exposition San Diego, California Jan.–March, 1912, Office of the Secretary Records, 1880–1929, Record Unit 45, Smithsonian Institution Archives, Washington, DC (hereafter OSR/SIA).
9. Ibid.
10. Ibid.
11. W. H. Holmes to Richard Rathbun, March 27, 1912, Box 22, Folder 5: Expositions: Panama California Exposition San Diego, California Jan.–March, 1912, OSR/SIA.
12. Aleš Hrdlička, "Advantages to the National Museum," memorandum attached to a letter from W. H. Holmes to Richard Rathbun, January 31, 1912, Box 22, Folder 5: Expositions: Panama California Exposition San Diego, California Jan.–March, 1912, OSR/SIA.
13. Aleš Hrdlička to W. H. Holmes, March 29, 1912, Box 22, Folder 5: Expositions: Panama California Exposition San Diego, California Jan.–March, 1912, OSR/SIA.
14. Rydell, *All the World's a Fair*, 220.
15. Ibid., 221.
16. D. C. Collier to Charles D. Wolcott, January 24, 1912, Correspondence 1912–1915, San Diego Exposition, Hrdlička Papers.
17. Rydell, *All the World's a Fair*, 221. The United States founded the Civil Government in the Philippines in 1901, and until the Woodrow Wilson administration, the United States controlled nearly all official affairs in the country. On the ground, this meant that those who were working on

behalf of museums and fairs in the country were given protection and military supervision, though this did not fully prevent danger from disease or backlash from indigenous peoples when they attempted to collect skeletons. In the case of Philip Newton, the U.S. Army offered to transport Newton and his collections to and from the Philippines; however, the Smithsonian ultimately decided to send Newton on a commercial ship. Newton's collections continued to be cared for by the army. C. D. Walcott to Quartermaster General, U.S. Army, May 21, 1912, Box 22, Folder 6: Expositions: Panama California Exposition San Diego, California April–June, 1912, OSR/SIA. On the decision to send Newton himself on a commercial ship, see C. D. Walcott to the Assistant Secretary of War, May 4, 1912, ibid.

18. Rydell, *All the World's a Fair*, 221.
19. Unknown author to Charles D. Walcott, June 14, 1912, Box 22, Folder 6: Expositions: Panama California Exposition San Diego, California April–June, 1912, OSR/SIA.
20. Aleš Hrdlička to W. H. Holmes, April 27, 1912, ibid.
21. Hiram Bingham to W. H. Holmes, January 7, 1913, Box 22, Folder 8: Exposition: Panama-California Expo, San Diego, Calif., 1913, OSR/SIA.
22. Hrdlička gathered over one thousand pathologic specimens, turning the collection over to the city of San Diego. In 1980, the San Diego Museum of Man published a catalogue of Hrdlička's original paleopathology collection. Spencer L. Rogers notes, "Dr. Hrdlička traveled to Peru and found that no excavation was needed as most of the ancient cemeteries had been despoiled by artifact hunters. Skulls and bones littered the surface of the sites. Because of this, he was able to examine an immense number of specimens and to make a selective collection in a relatively short time. The disadvantage of this was the lack of cultural association and disassociation of the bones." Spencer L. Rogers, foreword to *Catalogue of the Hrdlička Paleopathology Collection*, ed. Rose A. Tyson and Elizabeth Dyer S. Alcauskas (San Diego: San Diego Museum of Man, 1980), vii.
23. Aleš Hrdlička to W. H. Holmes, April 27, 1912, Box 22, Folder 6: Expositions: Panama California Exposition San Diego, California April–June, 1912, OSR/SIA.
24. C. D. Walcott to the Secretary of State, May 9, 1912, Box 22, Folder 7: Expositions: Panama California Exposition San Diego, California July–Dec. 1912, OSR/SIA.

25. Justo Perez Figuerola to the First Under-Secretary of the Ministry of Foreign Relations, June 21, 1912, ibid.
26. "Notes and Gleanings," *New York Times*, February 15, 1914, C4.
27. Hrdlička noted that his request to view the bones of *Pithecanthropus erectus*, or Java Man, was denied, and that this was happening "to everybody." Aleš Hrdlička to W. H. Holmes, October 19, 1912, Box 22, Folder 7: Expositions: Panama California Exposition San Diego, California July–Dec. 1912, OSR/SIA.
28. Aleš Hrdlička to W. H. Holmes, October 19, 1912, Box 22, Folder 7: Expositions: Panama California Exposition San Diego, California July–Dec. 1912, OSR/SIA.
29. Hrdlička's account of his travels appears in a letter written to W. H. Holmes, the long-standing head curator in the department of anthropology. In a lengthy thirteen-page report, Hrdlička describes both his success and his struggles in obtaining new information and collections. Aleš Hrdlička to W. H. Holmes, October 19, 1912, Box 22, Folder 7: Expositions: Panama California Exposition San Diego, California July–Dec. 1912, OSR/SIA.
30. Aleš Hrdlička to W. H. Holmes, September 12, 1913, Box 22, Folder 8: Expositions: Panama California Exposition San Diego, Calif., 1913, OSR/SIA.
31. Letter from Aleš Hrdlička to W.H. Holmes, January 31, 1914, Box 22, Folder 9: Exposition: Panama-California Exposition 1914–1916, OSR/SIA.
32. A. Schück to Aleš Hrdlička, September 9, 1914, ibid.
33. Aleš Hrdlička to W. H. Holmes, October 6, 1914, ibid.
34. Hrdlička, "Exhibit in Physical Anthropology," 408.
35. When the exposition opened, the laboratory was a working one, as planned. Workers soon found the room to be excruciatingly hot, with little air circulation. Edgar Hewett to Aleš Hrdlička, July 31, 1915, Hrdlička Papers.
36. Hrdlička, *Descriptive Catalog*, 7.
37. William Templeton Johnson, "San Diego: The Panama-California Exposition and the Changing Peoples of the Great Southwest," *Survey* 34, no. 14 (1915): 307, in Vertical File, Archives of the San Diego Museum of Man, San Diego, CA.
38. "Preparation of Exhibits Illustrating the Natural History of Man," 55–56.
39. Hrdlička, "Exhibit in Physical Anthropology," 409.
40. Hrdlička favored the term "Old Americans" to describe Caucasians whose families had been in North America for at least three generations. He also

used the term "thoroughbred," though he often placed the latter phrase in quotation marks. Hrdlička described the busts on one occasion using the following terms: "[The busts] are three series of true-to-nature busts, showing by definite age-stages, from birth onward and in both sexes, the three principal races of this country, namely, the 'thoroughbred' white American (for at least three generations in this continent on each parental side), the Indian, and the full-blood American negro." Smithsonian Institution, annual report, 1915, 11.

41. The Smithsonian noted, "The utmost care was exercised in ascertaining the age, particularly among the negro and Indian." "Preparation of Exhibits Illustrating the Natural History of Man," 59.
42. Ibid., 56–59.
43. Ales Hrdlička, "The Division of Physical Anthropology at the Panama-California Expos., San Diego," unpublished manuscript, 23, in Correspondence 1912–1915, San Diego Exposition, Hrdlička Papers.
44. Completing the displays in this room was a case comparing brains, skulls, and other bones at various stages in human aging. Hrdlička, "Exhibit in Physical Anthropology," 409–410.
45. Johnson, "San Diego," 307.
46. Hrdlička, "Exhibit in Physical Anthropology," 410.
47. "Preparation of Exhibits Illustrating the Natural History of Man," 59.
48. When Micka was hired, he was working under a sculptor in New York. Aleš Hrdlička to W. H. Holmes, March 21, 1912, Correspondence 1912–1915, San Diego Exposition, Hrdlička Papers.
49. Hrdlička, "Division of Physical Anthropology," 12–13.
50. "Preparation of Exhibits Illustrating the Natural History of Man," in Smithsonian Miscellaneous Collections, Vol. 65. Washington, DC, Smithsonian Institution, 1916, 59.
51. A smaller collection of Peruvian skulls showing evidence of trepanation was exhibited in 1893 at the World Columbian Exposition in Chicago. Following the fair, the private collection was transferred to the Mütter Museum in Philadelphia, where it remains on display today.
52. "Preparation of Exhibits Illustrating the Natural History of Man," 59.
53. Hrdlička, "Division of Physical Anthropology," 37.
54. Smithsonian Institution, annual report, 1915, 12.
55. In the 1970s, the San Diego Museum of Man received a grant from the National Science Foundation to make the original collections organized

by Hrdlička more accessible to the scholars and the public. Fliers advertising new publications intended for both teaching and research were distributed to museums and professionals, including the staff at the Mütter Museum. "Paleopathology for Teaching and Research," flier sent from the San Diego Museum of Man to Elizabeth Moyer, n.d., Folder: Bones, Curatorial Files, Mütter Museum, College of Physicians of Philadelphia, Philadelphia, PA.

56. Theodore Roosevelt to Aleš Hrdlička, March 1, 1915, Hrdlička Papers.
57. Aleš Hrdlička to Theodore Roosevelt, March 4, 1915, ibid.
58. James W. Wilkinson, "Exposition Excursions, Number Fourteen, Man's Evolution," newspaper clipping, *San Diego Union*, May 16 (probably 1915), Correspondence 1912–1915, San Diego Exposition, ibid.
59. Hrdlička, *Descriptive Catalog*, 11.
60. George G. Heye to Aleš Hrdlička, October 26, 1915, Hrdlička Papers. The manner in which complex ideas about race and evolution were broken down was also complimented by Charles Mayo, one of the founding physicians of the Mayo Clinic in Minnesota. See Edgar Hewett to Aleš Hrdlička, April 25, 1915, ibid.
61. Aleš Hrdlička to George G. Heye, November 2, 1915, ibid.
62. George G. Heye to Aleš Hrdlička, n.d., ibid.
63. Walter Hough, "History Co-worker with Anthropology," unpublished manuscript, Box 25-B, Folder: Manuscripts—Unpublished, Records of the Department of Anthropology, United States National Museum / National Museum of Natural History, Division of Ethnology, Manuscript and Pamphlet File, National Anthropological Archives, Smithsonian Institution, Washington, DC.
64. Ibid.
65. Joseph C. Thompson, "Savage Surgeons Fix Skulls: Crude Surgical Instruments Collected," newspaper clipping, *San Diego Union*, April 11, 1915, Correspondence 1912–1915, San Diego Exposition, Hrdlička Papers.
66. Sackman, *Wild Men*, 262–265.
67. Hrdlička, "Division of Physical Anthropology at the Panama-California Expos," 7. In the published catalogue for the exhibition, Hrdlička assures visitors, "These deficiencies, of which only the Preparator will be fully conscious, have already been partly compensated for and will further be done away with during the course of the Exposition." Hrdlička, *Descriptive Catalog*, 5.

68. The historian Constance Areson Clark documents how visual representations of evolution permeated society in the twentieth-century United States. Whereas many people in the United States proved to be trusting of science during the aftermath of the First World War, the decade of the 1920s was associated with a host of challenges to evolutionary concepts, evidence, she suggests, of an overriding fear of modernity. The use of visual illustration of evolutionary concepts at the 1915 fair and the many positive reactions to the displays provide a direct example of the popular acceptance of scientific ideas in the era of the First World War. Constance Areson Clark, *God—or Gorilla: Images of Evolution in the Jazz Age* (Baltimore: Johns Hopkins University Press, 2008).

69. A 1941 guidebook to Balboa Park describes the exhibitions after they moved to the larger California Building: "The Hall of Anthropology is the Museum's most notable department. It contains one of America's most important exhibits, tracing the development of the races of man. Included are replicas of prehistoric and historic racial types, their artifacts and tools, materials for comparative study of human anatomy with that of anthropoid après, and many charts and models illustrating racial differences and stages of the development in the human body. Also shown are notable skeletal materials revealing the diseases of prehistoric man, and methods of prehistoric surgery. An outstanding item of the exhibit is the collection of Peruvian trephined skulls, showing the trephine area and the method of bandaging." *A Guide to Balboa Park San Diego California*, American Guide Series (San Diego: Neyenesch, 1941), 17.

70. In chapter 6, I detail how the 1915 *Story of Man through the Ages* exhibit influenced the museum curator Henry Field in his creation of the 1933 *Hall of Prehistoric Man* at the Field Museum in Chicago. Henry Fairfield Osborn, a museum curator from the American Museum of Natural History, also used the production of the new images for a major publication. In his 1925 best-seller *Men of the Old Stone Age*, he borrowed images of busts originally produced by Rutot for the Panama-California Exposition. The historian Constance Clark does not note the fact that Rutot created the same busts for the 1915 exposition, but the statues are clearly reproduced from the originals created for *The Story of Man through the Ages* at the San Diego exposition. Clark, *God—or Gorilla*, 197.

5. SCIENTIFIC RACISM AND MUSEUM REMAINS

1. Circulated widely in the medical and anthropological communities of the era, this book is challenging to find today. A copy is in the W. Montague Cobb Papers at Howard University. W. Montague Cobb, *The Laboratory of Anatomy and Physical Anthropology of Howard University, 1932–1936* (Washington, DC, 1936), in Box 30, W. Montague Cobb Papers, Manuscript Division, Moorland-Spingarn Research Center, Howard University, Washington, DC (hereafter Cobb Papers).
2. Raymond Pearl, book review reprint, *Journal of Negro History*, October, 1936, in Box 34, Folder: Writings by Cobb—Reprints Book Reviews, ibid.
3. An important work documenting Howard University's influence on law, anthropology, and the construction of race is Lee D. Baker, *From Savage to Negro: Anthropology and the Construction of Race, 1896–1954* (Berkeley: University of California Press, 1998), 176–179. Baker notes that Cobb was joined on the faculty by the sociologist E. Franklin Frazier, the economist Abram L. Harris, and the philosopher Alain Locke. For additional context, see Elazar Barkan, *The Retreat of Scientific Racism: Changing Concepts of Race in Britain and the United States between the World Wars* (Cambridge: University of Cambridge Press, 1992).
4. Michael L. Blakey, "Cobb, William Montague (1904–1990)," in Spencer, *A History of Physical Anthropology*, 288.
5. Cobb was a prolific advocate for racial equality and the desegregation of medicine in the United States. One example includes W. Montague Cobb, *Progress and Portents for the Negro in Medicine* (New York: National Association for the Advancement of Colored People, 1948).
6. Cobb's collection of personal photographs even features snapshots of him attending events at the White House. These photographs feature Cobb alongside presidents John F. Kennedy and Lyndon B. Johnson. See Box 79, Photos—Framed, Cobb Papers. See also Box 77, Photos, ibid.
7. For more biographical information on Cobb, see Lesley M. Rankin-Hill and Michael L. Blakey, "W. Montague Cobb (1904–1990): Physical Anthropologist, Anatomist, and Activist," *American Anthropologist* 96, no. 1 (1994): 74–96.
8. Aleš Hrdlička to W. Montague Cobb, April 9, 1936, Correspondence, Cattell–Commerce Dept., Papers of Aleš Hrdlička, National Anthropological Archives, Smithsonian Institution, Washington, DC (hereafter Hrdlička Papers).

9. Quoted in Robert Oppenheim, "Revisiting Hrdlička and Boas: Asymmetries of Race and Anti-Imperialism in Interwar Anthropology," *American Anthropologist* 112, no. 1 (2010): 98. This quote was recorded while Hrdlička was serving as a witness at a congressional hearing on immigration.
10. Richard H. King, *Race, Culture, and the Intellectuals, 1940–1970* (Washington, DC: Woodrow Wilson Center Press; Baltimore: Johns Hopkins University Press, 2004), 1.
11. One of the most important books on the history of eugenics is Daniel J. Kevles, *In the Name of Eugenics: Genetics and the Uses of Human Heredity* (Cambridge, MA: Harvard University Press, 2004).
12. Alexandra Minna Stern, *Eugenic Nation: Faults and Frontiers of Better Breeding in Modern America* (Berkeley: University of California Press, 2005), 3.
13. A summary of Cobb's career and influence within the anthropological community can be found in Baker, *From Savage to Negro*, 184–186. Despite Baker's emphasis on Cobb and the Howard Circle, I would maintain that Cobb's role in the history of physical anthropology, medicine, and civil rights has been underemphasized in the literature overall.
14. While Boasian scholars like Alfred Kroeber shifted emphasis away from studying human remains, they still supervised growing collections of human remains in museums. Kroeber, as the curator of the Museum of Anthropology at the University of California, Berkeley (today the Hearst Museum of Anthropology), supervised a collection of human remains well on its way to becoming the largest collection of human remains in a museum of anthropology on the West Coast. While many of the remains were accessioned by other scholars working in the museum, including Edward Gifford and Robert Heizer, Kroeber's deemphasis of racial classification in his own work did not prevent him from allowing this work to continue at the museum he supervised.
15. Carleton S. Coon, *The Origin of Races* (New York: Knopf, 1962).
16. This topic is discussed in depth in the earlier chapters of this book; however, the best source on this tradition is Ann Fabian, *The Skull Collectors: Race, Science, and America's Unburied Dead* (Chicago: University of Chicago Press, 2010).
17. Clark Wissler, "Anthropological Collections in the American Museum of Natural History," unpublished manuscript, n.d., Correspondence:

Wineman-W3, Folder: Wissler, Clark, Hrdlička Papers. Although this document is undated, it appears to have been written about 1920.
18. Ibid.
19. Ales Hrdlička, *Anthropometry* (Philadelphia: Wistar Institute of Anatomy and Biology, 1920), 7.
20. Ibid.
21. Ibid., 8.
22. Ibid., 42.
23. Ibid.
24. Ibid., 44.
25. Ibid., 89.
26. Ibid.
27. Ibid., 101.
28. Ibid.
29. George Grant MacCurdy, "The First Season's Work of the School in France for Prehistoric Studies," *American Anthropologist* 24, no. 1 (1922): 61–71.
30. For more on the growth of anthropology in France, see Alice L. Conklin, *In the Museum of Man: Race, Anthropology, and Empire in France, 1850–1950* (Ithaca, NY: Cornell University Press, 2013).
31. A vast literature exists on the subject of the rise of museums in Europe, but especially in Britain, Germany, and France. Two examples of this literature are Philippa Levine, *The Amateur and the Professional: Antiquarians, Historians, and Archaeologists in Victorian England, 1838–1886* (Cambridge: Cambridge University Press, 1986); and Glenn H. Penny, *Objects of Culture: Ethnology and Ethnographic Museums in Imperial Germany* (Chapel Hill: University of North Carolina Press, 2007). For a general overview of the early development of a field of prehistory, see Donald R. Kelley, "The Rise of Prehistory," *Journal of World History* 14, no. 1 (2003): 17–36.
32. Ed Yastrow and Stephen E. Nash, "Henry Field, Collections, and Exhibit Development, 1926–1941," in *Curators, Collections, and Contexts: Anthropology at the Field Museum, 1893–2002*, ed. Stephen E. Nash and Gary M. Feinman (Chicago: Field Museum of Natural History, 2003), 127.
33. Affiliated with Yale's Peabody Museum between 1898 and 1931, MacCurdy authored several prominent texts and served in key leadership positions to expand the study of European prehistory and paleoanthropology in the

United States. George Grant MacCurdy, *Human Origins: A Manual of Prehistory*, 2 vols. (New York: D. Appleton, 1924).
34. Thomas C. Patterson, *Toward a Social History of Archaeology in the United States* (Fort Worth, TX: Harcourt Brace, 1995), 65.
35. MacCurdy, "First Season's Work," 63–65.
36. Ibid., 71.
37. Aleš Hrdlička to E. B. Renaud, May 14, 1923, Hrdlička Papers. Hrdlička suggests to Renaud, a professor at the University of Denver, that students prepare for the summer by reading Hrdlička's publication from the Smithsonian series "The Most Ancient Skeletal Remains of Man."
38. Charles Peabody to George MacCurdy, May 18, 1923, ibid.
39. Aleš Hrdlička to G. Steinmann, May 28, 1923, ibid.
40. Aleš Hrdlička to Charles Peabody, June 13, 1923, ibid.
41. Aleš Hrdlička to James Henry Breasted, June 19, 1922, Correspondence, Bohemian Circle-Breuil, ibid.
42. S. J. Redman, "What Self Respecting Museum Is without One? Collecting the Old World at the Science Museum of Minnesota, 1914–1988," *Collections: A Journal for Museum and Archive Professionals* 1, no. 4 (2005): 309–328.
43. Roland B. Dixon, *The Racial History of Man* (New York: Scribner, 1923).
44. Roland B. Dixon to Aleš Hrdlička, March 10, 1923, Hrdlička Papers.
45. Aleš Hrdlička to Roland B. Dixon, March 19, 1923, ibid.
46. Aleš Hrdlička, *Catalogue of Human Crania in the United States National Museum Collections* (Washington, DC: Government Printing Office, 1924).
47. Roland B. Dixon to Aleš Hrdlička, February 18, 1925, Hrdlička Papers.
48. Aleš Hrdlička to Roland B. Dixon, February 21, 1925, ibid.
49. Roland B. Dixon to Aleš Hrdlička, February 26, 1925, ibid.
50. Roland B. Dixon to Aleš Hrdlička, March 19, 1925, ibid.
51. Aleš Hrdlička to Roland B. Dixon, March 20, 1925, ibid.
52. Franz Boas, "Review of *The Racial History of Man* by Roland B. Dixon," *Science* 57, no. 1481 (1923): 587–590.
53. Thomas Wilson, "The Antiquity of Man in Its Relation to the Peopling of America: A Study of Prehistoric Anthropology," unpublished manuscript, n.d., in Box 85-A, Folder: Antiquity America—Wilson, Records of the Department of Anthropology, Division of Ethnology, Manuscript and Pamphlet File, National Anthropological Archives, Smithsonian Institution, Washington, DC.

54. Fay Cooper-Cole to Aleš Hrdlička, October 31, 1926, Correspondence, Cattell–Commerce Dept., Hrdlička Papers.
55. Though Cole is primarily remembered for his studies in cultural anthropology, he did have several students who entered the field of physical anthropology. One of Cole's students was Wilton M. Krogman, who became a prominent anthropologist in his own right. Krogman studied with Cole at Chicago before being instructed to study the human remains collections at the Field Museum of Natural History and Case Western Reserve University (the Hamann-Todd Osteological Collection, which was eventually turned over to the Cleveland Museum of Natural History).
56. Ben Hecht, *A Thousand and One Afternoons in Chicago* (New York: Covici, 1927), 65.
57. Montague W. Cobb, "Race and Runners," *Journal of Health and Physical Education* 7, no. 1 (1936), in Box 34, Folder: Writings by Cobb—R, Cobb Papers.
58. Montague W. Cobb, "The Physical Constitution of the American Negro," *Journal of Negro Education*, July 1934, 340, in Box 34, Folder: Writings by Cobb—The Physical Constitution–Pu, Cobb Papers. Cobb specifically points to the collections at Western Reserve University and Washington University, though he was certainly familiar with the collections at the Smithsonian Institution as well.
59. Ibid.
60. Ibid., 374.
61. Ibid., 374–375.
62. Ibid., 387.
63. Despite this tendency, Cobb did sometimes argue against the validity of race as an intellectual concept. In an article published in 1939, he wrote, "Against this distant and hopeful day when race will be a historical phase, others less auspicious must be anticipated." Montague W. Cobb, "The Negro as a Biological Element in the American Population," *Journal of Negro Education*, July, 1939, 347, in Box 34, Folder: Writings by Cobb—N.M.A–Negro, Cobb Papers.
64. T. D. Stewart, "The Growth of American Physical Anthropology between 1925 and 1975," *Anthropological Quarterly* 48, no. 3 (1975): 198–199.
65. Program of the Third Annual Meeting of the American Association of Physical Anthropologists held at the Smithsonian Institution, Washington, DC, March 21–23, 1932, in Hrdlička Papers.

66. Final Program of the Annual Meeting of American Association of Physical Anthropologists held at the Wistar Institute of Anatomy and Biology, Philadelphia, April 25–27, 1935, ibid.
67. Final Program of Annual Meeting, American Association of Physical Anthropologists, April 30–May 2, 1936, held at the Institute of Human Relations, Yale University, ibid.
68. Final Program of the Annual Meeting, American Association of Physical Anthropologists, April 8–10, 1937, held at Faculty Club Harvard University, ibid.
69. Cobb, *Laboratory of Anatomy and Physical Anthropology*, 1.
70. Ibid., 2.
71. Cobb considered the collection at the Hamann Museum of Comparative Anatomy and Anthropology to be the largest and best-documented collection of human remains to be found anywhere. The Hamann collection eventually became the basis for a museum of comparative anatomy. Following the Second World War, the collection was subsequently transferred to the Cleveland Museum of Natural History, where it is today known as the Todd-Hamann Collection. It remains one of the largest collections of human remains in the world.
72. Cobb, *Laboratory of Anatomy and Physical Anthropology*, 3.
73. Ibid.
74. Rather than taking a single tour of museums, Cobb describes his efforts along these lines as "visits to other departments whenever possible." He specifically notes visiting Harvard, Columbia, Pennsylvania, McGill, New York, and Washington Universities. Additionally, he describes visiting Harvard's group of natural history and anthropological museums, the Warren Anatomical Museum, the Boston Museum of Natural History, the American Museum of Natural History, and the Field Museum of Natural History. Finally, Cobb notes his visits to a series of zoos including those in New York, Philadelphia, St. Louis, Chicago, and Baltimore. Ibid., 7–8.
75. Ibid., 15–21.
76. Ibid., 45–48.
77. Ibid., 54.
78. Ibid., 54–55.
79. Ibid., 54.
80. Ibid., 57–58.
81. Ibid., 74.

82. Ibid.
83. Ibid., 74–75.
84. Ibid., 74–78.
85. For more on the place of the eugenics concept in the story of biological anthropology in the United States, see C. Loring Brace, *"Race" Is a Four-Letter Word: The Genesis of the Concept* (Oxford: Oxford University Press, 2005), 178–196.
86. Aleš Hrdlička to Rudolf C. Bertheau, January 25, 1940, Hrdlička Papers.
87. Quoted in Douglas Cazauz Sackman, *Wild Men: Ishi and Kroeber in the Wilderness of Modern America* (Oxford: Oxford University Press, 2010), 260–261. Franz Boas, like his student Kroeber, had directly attacked the eugenics movement in the early part of the twentieth century but received surprisingly little attention on the subject at the time. Baker, *From Savage to Negro*, 106–107.
88. Cobb was briefly on the board of directors of the AES in the early 1960s. Box 13, Folder: American Eugenics Society Inc., Cobb Papers. In his earlier work, Cobb described eugenics as "eugenical propaganda" and as "so dangerously liable to react unfavorably on minority groups that this approach is best left alone." Cobb, *Laboratory of Anatomy and Physical Anthropology*, 78.
89. In an important popular work originally published in 1944, the anthropologist Clyde Kluckhohn authored two consecutive chapters focusing on the practice of collecting skulls and skeletons in anthropology and the myth of race as a viable biological reality. The work represents just one of many in this vein but is representative of the turning of U.S. anthropology away from racial science toward other questions. Clyde Kluckhohn, *Mirror for Man: The Relation of Anthropology to Modern Life* (New York: Fawcett, 1944).
90. Franz Boas to Lucius N. Littaer, March 13, 1940, microfilm copy, Reel No. 43, Professional Correspondence of Franz Boas, American Philosophical Society, National Anthropological Archives, Smithsonian Institution, Washington, DC.

6. SKELETONS AND HUMAN PREHISTORY

1. Fay Cooper-Cole to Aleš Hrdlička, October 3, 1929, Correspondence, Cattell-Commerce Dept., Papers of Aleš Hrdlička, National Anthropological Archives, Smithsonian Institution, Washington, DC (hereafter Hrdlička Papers)

2. A summary of the manner in which ideas about evolution were challenged and displayed through visual images can be found in Constance Areson Clark, *God—or Gorilla: Images of Evolution in the Jazz Age* (Baltimore: Johns Hopkins University Press, 2008). For an overview of the early influence of Enlightenment-inspired, positivist thinking in the study of prehistory in the United States, see Alice B. Kehoe, *The Land of Prehistory: A Critical History of American Archaeology* (New York: Routledge, 1998), 64–96. Also illuminating the professionalization of archaeology in the United States during the era discussed in this chapter is Thomas C. Patterson, *Toward a Social History of Archaeology in the United States* (Fort Worth, TX: Harcourt Brace, 1995), 39–78.

3. For a lengthier meditation on this argument, see Gail Bederman, *Manliness and Civilization: A Cultural History of Gender and Race in the United States, 1880–1917* (Chicago: University of Chicago Press, 1995). Bederman's study concludes earlier chronologically than the 1933 fair, but the manner in which progress was viewed in relation to both evolution and history echoed throughout the ensuing decades.

4. T. D. Stewart, "The Growth of American Physical Anthropology between 1925 and 1975," *Anthropological Quarterly* 48, no. 3 (1975): 193–204.

5. Though not exhaustively, the question of the "museum period" and early museum culture in the United States has been explored by numerous historians in recent decades. Neil Harris, "The Gilded Age Revisited: Boston and the Museum Movement," *American Quarterly* 14, no. 4 (1962): 545–566. Included in a wave of new literature on museums was the work of George Stocking. See George Stocking, *Objects and Others: Essays on Museums and Material Culture* (Madison: University of Wisconsin Press, 1985); and George Stocking, *Victorian Anthropology* (New York: Free Press, 1987). One of the many historians of museums influenced by Stocking was Ira Jacknis. Critical to my own work are new case studies on the history of museum anthropology, including Ira Jacknis, "Alfred Kroeber as Museum Anthropologist," *Museum Anthropology* 17, no. 2 (1993): 27–32. Steven Conn further works to contextualize the historical trajectory of natural history museums in *Museums and American Intellectual Life, 1876–1926* (Chicago: University of Chicago Press, 1998).

6. In a book about the visual representation of ideas about human evolution, Constance Areson Clark vividly recounts the popular challenges faced by the American Museum of Natural History, as it mounted its own displays

on human evolution in the middle of the 1920s. Clark, *God—or Gorilla*, 25–27.
7. Ed Yastrow and Stephen E. Nash, "Henry Field, Collections, and Exhibit Development, 1926–1941," in *Curators, Collections, and Contexts: Anthropology at the Field Museum, 1893–2002*, ed. Stephen E. Nash and Gary M. Feinman (Chicago: Field Museum of Natural History, 2003), 137.
8. P. T. Barnum proposed a similar display of human remains in 1890. Barnum suggested that the mummified body of Ramses II could be acquired for a massive sum of $1 million. In an article circulated to newspapers and periodicals around the country, Barnum proposed that the corpse be displayed for visitors to the World Columbian Exposition, scheduled for three years later. Barnum died before the fair came to Chicago. "Barnum on the World's Fair," *Chicago Daily Tribune*, March 6, 1890, 9.
9. Fay Cooper-Cole to Aleš Hrdlička, October 3, 1929, Hrdlička Papers.
10. Ibid.
11. Ibid.
12. Henry Field to Aleš Hrdlička, July 15, 1927, Hrdlička Papers.
13. Field was related to Marshall Field, the wealthy department store owner and founder of the Field Museum, and Stanley Field, the president of the museum at the time.
14. Despite Field's rise to curator at the Field Museum of Natural History, he never became highly regarded within the broader anthropological community. Though his writings contributed somewhat to the discipline of physical anthropology, it was rarely groundbreaking, and many scholars suspected that his familial connections with Marshall Field helped Henry Field to secure a position as curator. Field eventually left the museum to assume a role in the navy during the Second World War. Mildred Trotter, an anatomist and physical anthropologist who studied at Oxford during the same period as Henry Field, argued in her oral history that Field had everything handed to him "on a silver platter." In particular, she resented the manner in which he was given his choice of research project, while she was assigned a particular project by their mutual advisers. Trotter seems to ascribe these events in part to Field's namesake and family wealth but also to her unique position as a woman in the male-dominated discipline of physical anthropology. Trotter's quote can be found in her unique oral history, Mildred Trotter, interviewed by Estelle Brodman, May 19

and 23, 1972, Becker Medical Library, Washington University, St. Louis, MO.

15. "I am leaving today to sail for Europe . . . to collect for the Hall of Prehistoric Man—which is to be exhibited here in the Museum at some future date." Field also notes that, by this point, the Field Museum had acquired about three thousand total sets of human remains. Field to Hrdlička, July 15, 1927.

16. Yastrow and Nash, "Henry Field, Collections, and Exhibit Development," 129–130. It might also be noted that the busts produced for the exhibition were reproduced in the best-selling publications on evolution by Henry Fairfield Osborn. Field, therefore, was not only influenced by the 1915 *Story of Man through the Ages* exhibit at the Panama-California Exposition in San Diego. Though the historian Constance Clark does not recognize the busts as such, they are clearly reproductions of the busts created for the 1915 displays. Clark, *God—or Gorilla*, 197.

17. Henry Field, "A Visit to Some California Museums March 15 to April 10, 1930," manuscript, Papers of Henry Field, Field Museum of Natural History Archives, Chicago, IL.

18. Ibid.

19. Yastrow and Nash, "Henry Field, Collections, and Exhibit Development," 129–130.

20. Aleš Hrdlička to Edgar L. Hewett, July 6, 1914, Hrdlička Papers.

21. Henry Field to Aleš Hrdlička, September 29, 1926, ibid.

22. "I have just returned to Chicago and hasten to write to thank you for all of your helpful suggestions for the Hall of Physical Anthropology." Henry Field to Aleš Hrdlička, January 10, 1930, ibid.; see also Roland B. Dixon to Aleš Hrdlička, March 19, 1925, ibid.

23. Yastrow and Nash, "Henry Field, Collections, and Exhibit Development," 130–131.

24. Berthold Laufer, preface to *The Races of Mankind, Sculptures by Malvina Hoffman: An Introduction to Chauncey Keep Memorial Hall*, by Henry Field, 3rd ed. (Chicago: Field Museum of Natural History, 1937), 4.

25. Arthur Keith, introduction to *Races of Mankind*, 7.

26. For more on Keith and his impact on American anthropology, including his influence on Earnest A. Hooton, see C. Loring Brace, *"Race" Is a Four-Letter Word: The Genesis of the Concept* (Oxford: Oxford University Press, 2005), 226–235.

27. Keith, introduction to *Races of Mankind*, 8.
28. Ibid., 13.
29. Ibid., 10.
30. While some of the rhetoric embedded in *The Hall of Races of Mankind* did discuss human evolution, it was largely focused on the differences between human types. *The Hall of Prehistoric Man* was far more concerned with explaining concepts of human evolution over the course of time.
31. Yastrow and Nash, "Henry Field, Collections, and Exhibit Development," 135–136.
32. "A Million Years of Man," *Science News-Letter* 24, no. 643 (1933): 85.
33. Henry Field, *On the Track of Man: Adventures of an Anthropologist* (Garden City, NY: Doubleday, 1953), quoted in Yastrow and Nash, "Henry Field, Collections, and Exhibit Development," 136–137.
34. "A Million Years of Man," 85, 87. Portions of this article (much of which may have simply been culled from a museum press release) were used in a similar article, "Evolution of Man Shown in Exhibits," *New York Times*, July 31, 1933, 15.
35. "Museum Tells Story of Man before History," *Chicago Daily Tribune*, July 31, 1933, 13.
36. Yastrow and Nash, "Henry Field, Collections, and Exhibit Development," 136–137.
37. In particular, the limited display of remains in these exhibitions might be contrasted to the displays from the 1893 World's Columbian Exposition and the 1915 Panama-California Exposition, both described in detail earlier in this book.
38. Herman J. Doepner to Aleš Hrdlička, April 9, 1927, Correspondence, Folder: DI-DOR (Miscellaneous) 1903–40, Hrdlička Papers.
39. Aleš Hrdlička to Herman J. Doepner, May 24, 1927, ibid.
40. Keith, introduction to *Races of Mankind*, 13.
41. Harvard University, annual report, 1932–1933, 301.
42. Franklin Roosevelt to Aleš Hrdlička, July 28, 1917, Hrdlička Papers.
43. Aleš Hrdlička to Franklin Roosevelt, February 25, 1933, ibid.
44. For more on Dart, see Janette Deacon, "Obituary: Raymond Arthur Dart (1893–1988)," *South African Archaeological Bulletin* 45, no. 151 (1990): 60.
45. Ann Gibbons, *The First Human: The Race to Discover Our Earliest Ancestors* (New York: Anchor Books, 2007), 6.

46. Aleš Hrdlička to Raymond Dart, February 21, 1925, Hrdlička Papers.
47. Raymond Dart to Aleš Hrdlička, January 4, 1925, ibid.
48. For a recent contribution to the global history of paleoanthropology in the twentieth century that sheds more light on Black in the context of his work with Chinese scientists, see Sigrid Schmalzer, *The People's Peking Man: Popular Science and Human Identity in Twentieth-Century China* (Chicago: University of Chicago Press, 2008). The book adds a different and complementary perspective on the discovery of Peking Man, exploring in-depth the significance of these and other fossils for Chinese society and culture. An earlier effort to historicize the story of Peking Man can be found in Harry L. Shapiro, *Peking Man* (New York: Simon and Schuster, 1975).
49. Gibbons, *First Human*, 35.
50. A letter from Aleš Hrdlička to Davidson Black offers an example of the slow and incomplete nature of academic communication during this period. Despite the fact that they were frequent correspondents and often exchanged reprints, Hrdlička wrote to Black, "Every now and then during the last year or so we have heard of 'marvellous' discoveries of ancient man in Mongolia or north-western China. Please tell me what is in it." Aleš Hrdlička to Davidson Black, October 10, 1925, Correspondence, Benes–Bogoras, Box 14 Hrdlička Papers.
51. Gibbons, *First Human*, 25–28.
52. In 1924, Hrdlička remarked, upon seeing a collection of original fossils collected by Eugène Dubois, that he was struck with how poorly the fossils had been represented by the casts he had been able to view and acquire in the United States. Smithsonian Institution, annual report, 1924, 10.
53. Edgar B. Howard to Henry B. Collins, Jr., December 31, 1936, Box 13, Incoming Letters Harper–Int., Folder: Holt–Huns, Papers of Henry B. Collins, National Anthropological Archives, Smithsonian Institution, Washington, DC (hereafter Collins Papers).
54. Ibid.
55. Arthur Keith to Aleš Hrdlička, January 30, 1935, Hrdlička Papers.
56. Aleš Hrdlička to Arthur Keith, February 11, 1935, ibid.
57. Arthur Keith to Aleš Hrdlička, February 13, 1936, ibid.
58. Kenneth A. R. Kennedy and T. Sheilagh Brooks, "Theodore D. McCown: A Perspective on a Physical Anthropologist," *Current Anthropology* 25, no. 1 (1984): 99–103.

59. Milford Wolpoff and Rachel Caspari, *Race and Human Evolution* (New York: Simon and Schuster, 1997), 274–275.
60. George Grant MacCurdy, ed., *Early Man: As Depicted by Leading Authorities at the International Symposium, the Academy of Natural Sciences, Philadelphia, March 1937* (Philadelphia: Lippincott, 1937).
61. Papers authored by Aleš Hrdlička, Herbert J. Spinden, Paul MacClintock, Ernst Antevs, Harold S. Gladwin, Kirk Bryan, and Frank H. H. Roberts, Jr., all address the problem of the peopling of the Americas. Only Hrdlička does so through a direct examination of human remains. See ibid.
62. This quote is from a speech that Hannah Marie Wormington delivered to the Philadelphia Congress of Anthropological and Ethnological Sciences in 1956. H. M. Wormington, "The Paleo-Indian," 1, in Box 5, Correspondence, Folder: Paleo-Indians, Papers of Hannah Marie Wormington, National Anthropological Archives, Smithsonian Institution, Washington, DC (hereafter Wormington Papers).
63. Scholars note that the Smithsonian's position on this unresolved debate had important influence as the field of archaeology professionalized in North America. Bruce G. Trigger, *A History of Archaeological Thought*, 2nd ed. (Cambridge: Cambridge University Press, 2006), 186–187.
64. Juliet Marie Burba, "Whence Came the American Indians? American Anthropologists and the Origins Question, 1880–1935" (Ph.D. diss., University of Minnesota, 2006),
65. J. D. Figgins, "The Antiquity of Man in America," *Natural History* 27, no. 3 (1927): 229, in Box 43, Miscellaneous, Accomplishments—East High School, Folder: Earliest Man in America [News Clippings], Wormington Papers.
66. "Although much has been learned about the implements of the Paleo-Indians we know little about the people who made them, for there has been a great death of ancient human bones." Wormington, "Paleo-Indian," 7.
67. Henry B. Collins, interviewed by Anna E. Riggs, August 2, 1978, 6, in Box 1, Articles on Collins about his work, 1927–1982, undated, Folder: "An Oral History Interview with Dr. Henry B. Collins," Collins Papers.
68. Ibid., 7.
69. Article reprint. J. D. Figgins, "The Antiquity of Man in America," reprinted from *Natural History* 27, no. 3 (1927): 229. In the Papers of Hannah Marie Wormington, Miscellaneous, Accomplishments-East High School, Folder:

Earliest Man in America [News Clippings], Box: 43, National Anthropological Archives. Smithsonian Institution.

70. Frank H. H. Roberts, Jr., "New World Man," *American Antiquity* 2, no. 3 (1937): 172–177.
71. At least one example from the Army Medical Museum collections points to the potential danger involved in collecting Native American human remains due to the possible violent reaction of nearby tribes. In one accession file, an army surgeon who collected a skull of an Ogallala Sioux woman describes the event as taking place "before the eyes of many Indian, who could see [him] in the distance"; he continues, "I had a lively adventure with it." Perhaps for this reason, he notes, he kept the skull for several years before submitting it to the Army Medical Museum. G. P. Hachenberg to Army Medical Museum, October 20, 1879, AMM #2034, Army Medical Museum Records, National Anthropological Archives, Smithsonian Institution, Washington, DC. David Hurst Thomas cites a letter from Franz Boas revealing feelings of guilt or remorse in desecrating graves. The letter was written while Boas was gathering skulls in British Columbia. On June 6, 1888, he wrote, "someone had stolen all the skulls, but we found a complete skeleton without head. I hope to get another one either today or tomorrow.... It is most unpleasant work to steal bones from a grave, but what is the use, someone has to do it." Quoted in David Hurst Thomas, *Skulls Wars: Kennewick Man, Archaeology, and the Battle for Native American Identity* (New York: Basic Books, 2000), 59.
72. Henry B. Collins, diary entry of July 29, 1929, Box 51, Diaries 1924–1930, Folder: Alaska 1929—typescript and carbon, 3/14–9/25, Collins Papers.
73. Aleš Hrdlička to George G. Heye, May 29, 1933, Hrdlička Papers.
74. George Heye to Aleš Hrdlička, May 31, 1933, ibid.
75. Melville J. Herskovits to Aleš Hrdlička, December 8, 1932, ibid. Herskovits became widely known for his work on African and African American studies. Herskovits wrote to Hrdlička about his attempts to collect skeletons in Africa just two years after the publication of the technical *Anthropometry of the American Negro* and four years after the appearance of the more popular book *The American Negro*. Melville J. Herskovits, *The Anthropometry of the American Negro* (New York: Columbia University Press, 1930); Melville J. Herskovits, *The American Negro: A Study in Racial Crossing* (New York: Knopf, 1928).

76. Kathleen S. Fine-Dare, *Grave Injustice: The American Indian Repatriation Movement and NAGPRA* (Lincoln: University of Nebraska Press, 2002).
77. Two sources provide the background information on Stewart in this paragraph. The first is a copy of his obituary, Eric Pace, "T. Dale Stewart Dies at 96: Anthropologist at Smithsonian," *New York Times*, October 30, 1997, 16. Also, detailed biographical information is included in Stewart's personal papers. Box 1, Folder: Biographical Information, Papers of T. Dale Stewart, National Anthropological Archives, Smithsonian Institution, Washington, DC (hereafter Stewart Papers).
78. T. Dale Stewart, interviewed by Pamela M. Henson, January–May 1975, 9, in Oral History Project Interviews, Archives and Special Collections of the Smithsonian Institution, Washington, DC.
79. Ibid., 15.
80. Ibid., 9–10.
81. Ibid., 31.
82. With contributions from the Peabody Museum at Harvard, the American Museum of Natural History, and the Smithsonian Institution, some early experiments with dried tissue samples proved encouraging, from the point of view of the scientists conducting the experiments. Ultimately, however, most of the attempts to restore dry tissue proved disappointing, with most connective tissue and internal organs badly deteriorated in most cases. Harris Hawthorne Wilder, "The Restoration of Dried Tissues, with Especial Reference to Human Remains," *American Anthropologist* 6, no. 1 (1904): 1–17.
83. Basic biographical information and some detail about the course of Trotter's life and career can be found in her oral history. Mildred Trotter, interviewed by Estelle Brodman, May 19 and 23, 1972, Becker Medical Library, Washington University, St. Louis, MO.
84. George Woodbury and Edna T. Woodbury, *Differences between Certain of the North American Indian Tribes as Shown by a Microscopical Study of Their Head Hair* (Denver, CO: State Museum, 1932).
85. Mildred Trotter, "Hair from Paracas Indian Mummies," *American Journal of Physical Anthropology* 1 (1943): 69–75.
86. William C. Boyd to T. D. Stewart, October 19, 1942, Box 3, Folder: Boyd, William C., Correspondence, ca. 1935–1961, Series 1, A–Boyd. Stewart Papers.
87. T. D. Stewart to William C. Boyd, September 26, 1939, ibid.

88. William C. Boyd to T. D. Stewart, September 26, 1939, ibid.
89. "I think a good deal of your work and wish you all possible success." Aleš Hrdlička to William C. Boyd, September 20, 1934, Correspondence, Bohemian Circle–Breuil, Hrdlička Papers.
90. William C. Boyd to Aleš Hrdlička, February 1, 1934, ibid.
91. Stewart, "Growth of American Physical Anthropology," 193.
92. Ibid., 196.
93. Ibid., 202. Stewart cites Clark Wissler, "The American Indian and the American Philosophical Society," *Proceedings of the American Philosophical Society* 86 (1943): 189–204.
94. Edgar Hewett to Aleš Hrdlička, October 16, 1934, Hrdlička Papers.
95. Ibid.
96. Aleš Hrdlička to Edgar Hewett, October 22, 1934, ibid.
97. Aleš Hrdlička to Arthur Keith, January 3, 1938, ibid.
98. Stewart, "Growth of American Physical Anthropology," 202–203.
99. Sherwood Washburn to Frank Spencer, June 15, 1983, Box 4, Folder: Washburn, Sherwood, Series 1, Correspondence, Papers of Frank Spencer, National Anthropological Archives, Smithsonian Institution, Washington, DC (hereafter Spencer Papers).
100. "Student of Man," *New York Times*, September 7, 1943, 22.
101. Aleš Hrdlička to James H. Brested, October 19, 1933, Correspondence, Bohemian Circle–Breuil, Hrdlička Papers.
102. The appeal for more cooperation in anthropology did not start with Hrdlička in the 1930s. The significance of this rhetoric is perhaps most appropriately understood as a desire for physical anthropology not to be left behind by other branches of the field that were becoming increasingly centered on university-based work. Fay Cooper-Cole, for instance, in a letter written to Hrdlička in 1926 called on the branches of anthropology to work together: "I feel that Anthropology is now in a position where co-operation of the workers is all that is needed to see our case materially advanced." Fay Cooper-Cole to Aleš Hrdlička, December 7, 1926, Correspondence, Cattell–Commerce Dept., Hrdlička Papers.
103. Melville J. Herskovits to T. D. Stewart, April 6, 1943, Box 7, Correspondence, ca. 1935–1961, Series 1, Gregory–Howell, Stewart Papers.
104. "Editorial," *American Journal of Physical Anthropology* 1, no. 1 (1943): 1.
105. Ibid., 2.
106. Ibid., 3.

107. J. Alden Mason to T. D. Stewart, November 26, 1945, Box 1, Folder: Publications—Anthropology, American Anthropologist, Biographical Material, Correspondence, Professional Publications, Stewart Papers.
108. T. D. Stewart to J. Alden Mason, November 15, 1945, ibid.
109. S. L. Washburn to W. D. Strong, March 7, 1945, Box 15, Folder: W–WA, Correspondence—Strong W.D.–Wei, Papers of William Duncan Strong, National Anthropological Archives, Smithsonian Institution, Washington, DC.
110. "Rise of Man," script, November 8, 1936, *The World Is Yours* Radio Program, Accession 06-152, Box 1, Smithsonian Institution Archives, Washington, DC. Some sources refer to this episode with the title "Early Man," while others refer to it as "Rise of Man."
111. Smithsonian Institution, annual report, 1937, 19–21. And Smithsonian Institution, annual report, 1938, 9–10.
112. Aleš Hrdlička, unpublished essay, 1943, 15, in Box 156, Manuscripts of Writings, 1941–1944, Folder: History of Physical Anthropology in the USA w/ Special Reference to Phila., Hrdlička Papers.
113. Sherwood Washburn to Frank Spencer, June 15, 1983, Box 4, Folder: Washburn, Sherwood, Series 1, Correspondence, Spencer Papers.
114. Lee D. Baker, *From Savage to Negro: Anthropology and the Construction of Race, 1896–1954* (Berkeley: University of California Press, 1998), 209–210.
115. Russell H. Tuttle, "Five Decades of Physical Anthropology," *Science* 220, no. 4599 (1983): 832–834.
116. Ashley M. F. Montague, *An Introduction to Physical Anthropology* (Springfield, IL: Charles C. Thomas), 103.
117. A prominent example of an anthropologist pointing to racial divisions was Earnest Hooton, who wrote, "Differences between Negroes and Whites, or between Mongoloids and Whites or Negroes, are really much greater than 'racial' differences; they are almost, if not quite, sub-specific or specific differences. Races are properly physical subdivisions of these three great general groupings of modern man." Despite the lingering prevalence of this idea, this type of argument was gradually sliding out of mainstream science. Earnest A. Hooton, *Apes, Men, and Morons* (New York: Putnam, 1937), 128.

EPILOGUE

1. Stephanie A. Makseyn-Kelley and Erica Bubnaik Jones, "Inventory and Assessment of Human Remains from the Historic Period Potentially Affiliated with the Eastern Dakota in the National Museum of Natural History," memorandum, April 24, 1996, Repatriation Office, National Museum of Natural History, Smithsonian Institution, Washington, DC.
2. The literature on NAGPRA has grown extensively in recent years. The following is a partial list of recent publications and reports on the subject: Sangita Chari and Jamie M. N. Lavallee, eds., *Accomplishing NAGPRA: Perspectives on the Intent, Impact, and Future of the Native American Graves Protection and Repatriation Act* (Corvallis: Oregon State University Press, 2013); Kathleen S. Fine-Dare, *Grave Injustice: The American Indian Repatriation Movement and NAGPRA* (Lincoln: University of Nebraska Press, 2002); Devon Abbott Mihesuah, ed., *Repatriation Reader: Who Owns American Indian Remains?* (Lincoln: University of Nebraska Press, 2000); Sangita Chari and Lauren A. Trice, *Journeys to Repatriation: 15 Years of NAGPRA Grants (1994–2008)* (Washington, DC: National Park Service, U.S. Department of Interior, 2009); Senate Committee on Indian Affairs, *Finding Our Way Home: Achieving the Policy Goals of NAGPRA: Hearing before the Committee on Indian Affairs*, 112th Cong., 1st sess. (June 16, 2011).
3. House Committee on Natural Resources, *Native American Graves Protection and Repatriation Act: Hearing before the Committee on Natural Resources*, oversight hearing, 111th Cong., 1st sess. (October 7, 2009).
4. National Museum of Natural History, Repatriation Office, "Frequently Asked Questions," http://anthropology.si.edu/repatriation/faq/index.htm#collect03 (accessed December 1, 2013).
5. David Hurst Thomas, "Repatriation: The Bitter End or a Fresh Beginning?," *Museum Anthropology* 15, no. 1 (1991): 10.
6. Samuel J. Redman, "Smithsonian Embarks on Move of Human Remains Collections," *Anthropology News* 52, no. 1 (2011): 22.
7. Kim Tallbear, *Native American DNA: Tribal Belonging and the False Promise of Genetic Science* (Minneapolis: University of Minnesota Press, 2013).
8. Smithsonian Institution, "Last Chance to Visit 'Written in Bone: Forensic Files of the 17th-Century Chesapeake," press release, July 31, 2013, http://newsdesk.si.edu/releases/last-chance-visit-written-bone-forensic-files-17th-century-chesapeake.

9. On the reorientation of museums in the second half of the twentieth century, see Stephen E. Weil, "From Being about Something to Being for Somebody: The Ongoing Transformation of the American Museum," *Daedalus* 128, no. 3 (1999): 229–258.
10. Randi Korn & Associates, Inc., "Responses to a Human Remains Collection: Findings from Interviews and Focus Groups," prepared for the National Museum of Health and Medicine, July 1999.

ACKNOWLEDGMENTS

Early in my studies at the University of Minnesota, Morris, my mentors began to introduce me to the complex history surrounding anthropology and archaeology in North America. At the time, I was unaware that this would become an intellectual journey lasting more than a decade. Years of reading, traveling to archives, and endless revising have now resulted in this book. Early in the process, I was supremely fortunate to start my professional career, while still a student, as an intern working at the Field Museum of Natural History in Chicago. This proved to be my first true exposure to the problem and potential of human remains in museums.

Later I moved to Denver, where I had the distinct honor of playing a small role in one of the largest repatriations and reburials in recent history, in 2006, near the spectacular Mesa Verde. I witnessed firsthand as twenty-four tribal groups came together with the National Park Service and Colorado History Museum to return ancient ancestors to the earth. These experiences left me curious, unsettled, and interested to learn more about the story behind these collections. How and why did so many museums in the United States end up acquiring thousands of bodies? Given the centrality of death and burial in the human experience, how could seemingly sacred principles be violated so directly and systematically?

Many people helped me turn this book into a reality, and I owe them an enormous debt of gratitude. Notably, I owe a special thanks to many individuals at the Smithsonian Institution, including JoAllyn Archambault, Alan Bain, William Billeck, Stephanie Christensen, Joanna Cohan Scherer, James Deutsch, Leanda Gahegan, R. Eric Hollinger, Jake Homiak, David Hunt, Lars Krutak, Robert Leopold, Daisy Njoku, Michael Pahn, Gina Rappaport, David Rosenthal, Jennifer Snyder, and Lorain Wang. Thanks to Pamela M. Henson and Ellen Alers at the Smithsonian Institution Archives.

At the University of Pennsylvania's University Museum Archives, I would like to thank Alessandro Pezzati. At the Phoebe A. Hearst Museum of Anthropology, I would like to especially thank the scholar Ira Jacknis and the archivist Joan Knudson. Thanks to Richard Wilshusen and Bridget Ambler, who generously shared ideas and resources on the history of archaeology connected to the Colorado Historical Society, now History Colorado. At the San Diego Museum of Man, special thanks to Tori Randall, Kelly Revak, and Sara Pianavilla. The entire staff at the Howard University Archives proved helpful and positive, most recently Kenvi Phillips, who assisted me with records and images related to W. Montague Cobb. Thanks to Armand Esai and Lauren Hancock at the Field Museum of Natural History.

These individuals contributed to this book by helping locate unique records, sharing ideas, and offering feedback on the project as it evolved. Archivists are the unsung heroes in our field, and I am grateful for having worked with such tremendous people.

I am exceedingly grateful to several organizations that helped make this book possible through institutional or financial support. My first year in the archives at the Smithsonian was made possible by a grant from the Center for Race and Gender and a Graduate

Division Summer Grant, both from the University of California, Berkeley. Subsequent support from the Department of History at Berkeley helped me continue my research in the archives. Following my arrival at the University of Massachusetts, Amherst, I was generously provided with research support from the College of Humanities and Fine Arts as well as the Massachusetts Society of Professors. I am fortunate to have great colleagues as mentors at UMass who graciously offered guidance on this project.

At the Mütter Museum and College of Physicians of Philadelphia, I would like to thank Sofie Sereda, Annie Brogan, Anna Dhody, and Robert Hicks. A portion of the research for this project was funded by a pair of F. C. Wood Institute for the History of Medicine Travel Grants from the College of Physicians of Philadelphia. My time at the Mütter Museum and College of Physicians Library provided me with the tools to better understand the foundation of medical museums in the United States.

My wonderful mentors from the History Department at the University of California, Berkeley, included Richard Cándida Smith and Randolph Starn. In the Ethnic Studies Department, I owe thanks to Thomas Biolsi. Thanks to all three scholars for their ongoing mentorship, support, and guidance. Thanks also to my undergraduate mentors Roland Guyotte and Julie Pelletier, who fostered my interest in this topic from the earliest stages.

This book would not have been possible without Diana Tung, Len Hamilton and Robin Timmons, Emily Esten, and the anonymous reviewers who helped to improve this project through careful reading and thoughtful feedback.

My tremendous editor at Harvard University Press, Kathleen McDermott, helped take this project to the next level. I am eternally grateful for her steadfast support, feedback, and consistent guidance.

I dedicate this book to my family and friends. Important in supporting me in writing this book were my beautiful wife, Emily, whose historian's eye and gourmet cooking nourished both the manuscript and me; my sister and friend, Elisa; and my amazing mother, Cindy Redman. Most especially, I dedicate this book to my late father, Jim Redman, who supported both the project and me from the start.

INDEX

Academy of Natural Sciences, Philadelphia: American Giant and, 140; International Symposium on Early Man at, 248–251
Aetas, 78
Africa: collecting in and specimens from, 27, 159, 171, 256, 269–270; evolution and, 200, 240, 246–249. *See also* British East Africa; South Africa
African Americans: athletic ability and, 211–212; collecting of remains of, 15, 27, 30, 32; Morton's racial theories and, 25; physical constitution of, 212–215, 339nn55,63. *See also* Howard University; Racial classification; Scientific racism
Agassiz, Louis, 300n17
Age and aging: American Association of Physical Anthropology and, 215–218; anthropometric measurements and, 196; shown in *Story of Man through the Ages*, 176–178, 180–181, 186–187
Agnew, D. Hayes, 143
Alaska: collecting in and specimens from, 18–20, 103, 164, 168, 252–253, 255, 258–259; NAGPRA and, 279; prehistory and, 252–253
Aleutian Islands, 18, 255–256
Allen, J. F., 131
All the World's a Fair: Visions of Empire at American International Expositions, 1876–1916 (Rydell), 328n1
Amateur collectors, 6, 15, 18, 20, 42–43, 56, 66, 70–71, 73–75, 86, 109, 124–125, 266. *See also* Looters and looting
American Anthropological Association (AAA), 264, 314n36
American Anthropologist, 58–59, 97, 271

American Antiquities Act (1906), 73–74, 75, 106, 162, 299n9; intentions and results of, 73–74, 83–91
American Association for the Advancement of Science (AAAS), 22, 47, 313n22, 314n36
American Association of Physical Anthropologists (AAPA), 189, 215–218, 271
American Eugenics Society (AES), 55, 118, 122, 223, 224, 341n88
American exceptionalism, 120–121
American Giant, at Mütter Museum, 140–141, 149
American Journal of Physical Anthropology, 224, 246, 261; Hrdlička and, 54, 118, 250–251; Stewart and, 259, 264, 270–271
American Medical Association (AMA), 152
American Museum of Natural History (AMNH), 46, 55, 58, 63, 97, 225, 298n35; eugenics and, 119, 121–122; Magdalenian Girl and, 231; physical anthropology and, 193–194, 264–265
American Naturalist, 62–63
American School in France for Prehistoric Studies (ASFPS), 198–201
American Southwest. *See* Cliff dwellers, of American Southwest
American Surgical Association (ASA), 325n54
Amsterdam International Colonial Exposition (1883), 310n119

Anatomy Act of Pennsylvania (1867), 130, 322n14
Ancestry, 60, 217; eugenics and, 122; Hrdlička and, 54–55, 197–198; racial classification and, 60–61; racial history and, 74
Ancient history. *See* Prehistory and human origins
Angel, Lawrence, 258
Anthropological Quarterly, 229–230
Anthropological Society of Washington, 22
Anthropometric measurements, 136; Boas and, 45–46; eugenics and, 120–122; Hrdlička and, 80, 93, 194–198, 326n68; medical history and, 196; taken at expositions, 160, 172, 244; taken at museums, 104–105
Antiquities: collecting difficulties and, 166–167; uses of term, 166
Archaeological Institute of America, 22–23, 314n36
Army Medical Museum (AMM), 106, 256, 286; call for military materials, 28–34, 130, 302n33; comparative anatomy and, 5, 26–27, 33–34, 259; cooperation with Mütter Museum, 146; dangers of collecting and, 348n71; as early collector of skeletons, 18; establishment of, 26; ethnographic materials transferred to Smithsonian, 28, 34–36, 51–53, 76, 129, 303nn53,54,55; lost limbs of veterans and, 131–133; at Louisiana Purchase Exposition, 137; medical

education and, 29, 32, 302n35; move to Ford's Theater, 27, 31, 131; pathology focus of, 128–129, 321n6; public interest in displays at, 27–28, 133, 301n30; Sioux man's remains and, 3, 4–5; soft-tissue samples and, 136. See also National Museum of Health and Medicine
Army Medical School, 131
Artistic renderings (busts, casts, masks), used in Story of Man through the Ages, 159, 161, 165, 169–170, 172–180, 183, 186–187, 334n68
Ashberry, Mary, 324n50
Athletic ability, of African Americans, 211–212
Auditory apparatus, collection of, 135, 148
Australopithecus afarensis, 217

Bandelier, Adolph F., 22, 47, 306n81, 313n29
Barnum, P. T., 36–37, 232, 343n8
Barrett, Samuel A., 79
Basket Maker, blood and, 262–263
Beam, George L., 90
Bell, Whitfield J. Jr., 321n9
Bering Strait land bridge, 170, 173, 249
Biblical narratives, ancient history and, 201–203
Bingham, Hiram, 166–167
Biocolonialism, 281
Black, Davidson, 247, 346nn48,50
Blood types, racial classification and prehistory, 259, 262–263

Boas, Frank, 10, 264; American Antiquities Act and, 85–86; background and training of, 45–46; criticism of Dixon's racial stability notion, 206–208; cultural anthropology and, 191–192, 225–226, 336n14; environment and, 120–121; eugenics and, 223–224, 341n87; exhibitions and research at World's Columbian Exposition, 44–45, 46–47, 305n74; Hrdlička and, 185–186; as "primary collector," 109; racial classification and, 58–62, 123, 215–216; remains collected by, 66; remorse over collecting practices, 348n71; as teacher, 94, 315n52
Body Worlds exhibitions, 288
Bokovoy, Matthew F., 328n1
Bolivia, 193
Bone rooms, establishment of, 3–4
Boyd, William C., 262
Brace, C. Loring, 300n17
Breasted, James H., 202
Brinton, John Hill, 28
British Columbia, 46, 305n71, 348n71
British East Africa, 170–171
British Medical Journal, 152, 153
British Museum, 199
Broca, Paul, 56
Broom, Robert, 217, 248, 251
Brown, Julie K., 134
Buffalo, NY, exposition (1901), 64
Bullet wounds, samples of, 136
Bunker, Chang and Eng, 138–140, 141

Bureau of American Ethnology
(BAE), 21, 28, 34, 49, 75–76, 149
Busts. *See* Artistic renderings (busts, casts, masks)

Cap Blanc skeleton, 210, 231
Carnegie, Andrew, 152–153
"Case of George Dedlow, The" (Mitchell), 132–133
Case Western Reserve University, 214, 218, 339n55, 339n58
Casts. *See* Artistic renderings (busts, casts, masks)
Centennial exhibition, in Philadelphia (1876), 19, 64, 135, 136
Central Eskimo, The (Boas), 45
Century of Progress International Exposition (1933), 227, 229, 234, 244
Chapin, F. H., 47
Chicago Daily Tribune, 78, 242
China, 168, 245, 247
Chippewa Indians, Densmore's discoveries of artifacts of, 69–72
Civil War, 28–34, 131–133, 302n33
Clark, Constance Areson, 334n68, 344n16
Cleveland Museum of Natural History, 339n55, 340n71
Cliff dwellers, of American Southwest: exhibitions and, 42–44, 50–51, 304n66; impact on later federal laws, 42–43; in museums, 66–67; public interest in, 16–21, 39–40; racial classification and, 38–39, 41–42, 304n61; World's Columbia Exhibition and, 47–50, 135–136

Clovis, NM, 252
Cobb, W. Montague, 282, 339n63; athletic ability and race and, 211–212; eugenics and, 224, 341n88; medical education and, 188–191, 218–223, 340nn71,74; racial classification and, 215–216, 225
Collections and collecting, 13–14, 73–74; 117–123; American Antiquities Act and, 73–74, 83–91; contemporary display practices and new scientific study techniques, 280–284, 287–288; early bone rooms and, 3–4; ethical concerns, in twenty-first century, 279, 282–283, 285–290; ethnographic professionals and, 108–115, 124; eugenics and, 75, 117–123, 124; growth of, 6–7; illicit acquisitions, 106–108; Ishi and his brain, 94–101, 316n64; nationalism and, 198–201, 202; prehistory study, 74–75, 101–104; racial questions and disarray for collections, 75–83; repatriation programs and, 277–280, 289; risks of, 29; salvage anthropology and, 69–73, 123; specimen preservation and, 126, 154, 321n3; storage and, 104–105, 123–124; for *Story of Man through the Ages*, 165–171; tourists and, 27, 74, 85–87, 90, 133; valuing and exchanging of objects, 115–117; World War I trenches and "war opportunity," 91–94

College of Physicians, Philadelphia, 129, 130, 141, 149, 321n9. *See also* Mütter Museum, Philadelphia
Collins, Henry Bascom, 253, 254, 255
Colorado Museum of Natural History, 252, 254
Comparative anatomy: Army Medical Museum and, 5, 26–27, 33–34, 136, 259; cliff dwellers and, 17
Congress of Hygiene and Demography, 117
Conjoined twins, 138–140, 141
Conn, Steven, 7, 296n22
Coon, Carleton S., 192
Cooper-Cole, Fay, 209, 227–229, 232, 339n55, 350n102
Corbusier, William Henry, 31–32, 302n45
Cradleboarding, of Native American infants, 150–151
Crania Americana (Morton), 23
Crania Ethnica America (Virchow), 327n75
Cultural anthropology: Boas and, 191–192, 225–226, 264, 336n14; eugenics and, 223–224; Herskovits and, 264, 269–270
Cushing, Frank Hamilton, 49, 83
Cuvier, Georges, 54

Da Costa, J. M., 143
Dakota man. *See* Sioux Indian
Dart, Raymond, 246–247
Darwin, Charles, 26, 268, 301n29
Darwinian natural selection, 57
Delight Makers, The (Bandelier), 47, 306n81, 313n29

Densmore, Frances, 69–72, 108–110, 124
Denver and Rio Grande Railroad, 90
Dioramas, used at Field Museum exhibits, 234, 240–242
Dixon, Roland B., 204–209, 225
Doepner, Herman J., 243
Dorsey, George, 109
Dubois, Eugène, 246, 251, 346n52
Durango, CO, mummies exhibited in, 43

Egypt: evolution and, 240; mummies from, 37, 38, 40, 194, 203, 257, 262–263; public interest in, 17, 36, 230
Environment, human evolution and, 76–78, 89, 120–121, 148–149
Eskimos: crania collected, 193–194; at Field Museum, 116; Kishu, 97, 316n57; prehistory and, 252–253; at St. Louis World's Fair, 64–65, 310n22
Ethical concerns: in twenty-first century, 10–11, 279, 282–283, 285–290; of acquisitions and displays, 13–15, 70, 77, 107–108, 135, 239, 254–257
Ethnographers, as collectors, 108–115, 124
Ethnomusicologists, 70, 108
Eugenics, 55, 134, 146, 208; "hygiene" and, 137, 319n98; influence on collecting and collections, 117–123; scientific racism and, 6, 191, 192, 223–225, 285–286, 341nn87,88; *Story of Man through the Ages* and, 182–183

Evolution: environment and, 76–78, 89, 120–121, 148–149; eugenics and, 121–122; Hrdlička and, 82; physical anthropology and, 274–275; prehistory and, 246–251; racial classification and, 56–61
Exposition Universelle in Paris (1889), 310n119

Field, Henry, 116, 199; Field Museum's 1933 exhibits and, 228–229, 244, 334n70; Smithsonian and, 237; *Story of Man*'s influence on, 228–229, 234–236, 344n16
Field Museum of Natural History, 244, 298n35, 339n55; exhibits of race and prehistory, generally, 228–230, 232–244, 274, 334n70; *Hall of Prehistoric Man* exhibit, 234, 240–242, 249, 334n70, 344n15, 345n30; *Hall of Races of Mankind* exhibit, 220, 231, 236–240, 242, 244, 249, 345n30
Figgins, J. D., 252, 253 254
Fine-Dare, Kathleen, 297n24
Finns, athletic ability of, 211–212
First Nations people, of Canada, 45, 305n71
Fletcher, Alice, 85, 89, 109, 110–112, 124, 318n86
Folsom, NM, 251–252
Ford's Theater 27, 131
Forwood, W. H., 30
France, war opportunity and, 92–93
Funerary objects, number of remains still in museums, 15, 298n34

Gender: anthropometric measurements and, 196; in Army Medical Museum–Smithsonian collection transfer, 35, 303n55; organization of collections and, 8, 62–63
Genetics, increased research reliance on, 280–281
Georgetown College, 32
Gifford, Edward, 336n14
Gordon, Ed, 211
Gould, Stephen J., 24
Grave robbing. *See* Looters and looting

Hair, racial classification and prehistory, 259–261, 263
Hall of Prehistoric Man exhibit, at Field Museum, 234, 240–242, 249, 334n70, 344n15, 345n30
Hall of Races of Mankind exhibit, at Field Museum, 220, 231, 236–240, 242, 244, 249, 345n30
Hamann Museum of Comparative Anatomy and Anthropology, 340n71
Hamann-Todd Osteological Collection, 339n55, 340n71
Hammond, William, 26
Harper's Magazine, 40, 304n61
Harrington, John P., 109, 113, 124
Harvard University: Peabody Museum at, 244, 298n35; Warren Anatomical Museum at, 320n1
Haskell, Thomas, 296n21
Hawaii, 279
Hazzard, C. D., 48–49, 307nn85,87

Hecht, Ben, 210
Heizer, Robert, 315n53, 336n14
Heredity. *See* Ancestry
Herskovits, Melville J., 256, 264, 269–270, 348n75
Hewett, Edgar Lee: American Antiquities Act and, 88–89, 314nn36,37,38; Panama-California Exposition and, 162–163, 238–239; physical anthropology and, 265–266; *Story of Man through the Ages*, 234, 265; World War I and, 91–92
Heye, George Gustov, 112, 114, 183–184, 255–256, 318n91
Heye Museum, 112
Hoffman, Malvina, 237
Holmes, William Henry, 34–35, 88, 92, 98, 102, 252; Panama-California Exposition and, 162–163, 166–167
Homo erectus, 240, 247
Homo novusmundus, 254
Hooten, Earnest Albert, 10, 54–55, 351n117
Hough, Walter, 184–185
Howard University, Cobb and medical education at, 188–191, 218–223, 282
Hrdlička, Aleš, 10, 66, 71, 185, 202, 209, 251, 300n12, 346nn50,52; African Americans' physical constitutions and, 213, 214; Alaska and, 253, 255–256; American Museum of Natural History and, 193; American School in France for Prehistoric Studies and, 200–201; anthropometry and, 80, 93, 104–105, 136, 326n68; background of, 53, 55; classification and, 62–63; on Cobb, 189–190; collection exchanges and, 115–116; complex legacy of, 53–56, 264, 267–268, 280, 282, 283–284; criticism of Dixon's methodological approach, 204–206, 208; Dart and, 246, 247; dating human arrivals in North America and, 252, 254; declines to work on Field Museum exhibits, 232–233; Eskimos at St. Louis World's Fair and, 64–65, 310n22; eugenics and, 55, 117–123, 223; Field and, 237; as first Smithsonian curator, 53, 56; guide to acquisitions, 67–68; guide to collections, 76–78; guide to taking anthropometric measurements, 194–198; Herskovits on, 270; Heye and, 112, 114, 318n91; on importance of studying remains, 72; Ishi and, 97, 98, 101; Keith and, 250; McCown and, 250; organization and purpose of Smithsonian collections, 75–83; physical anthropology and, 263, 264–268, 269, 273–274, 350n102; prehistory and, 102–103; as "primary collector," 109; racial classification and, 58, 61, 243–244; Roosevelt and Japanese race and, 244–245; Stewart and, 258–259; Stewart contrasted, 257–258; war opportunity and, 91–93

Hrdlička, Aleš, and *Story of Man* exhibition, 238–239, 265; correspondence about, 168, 181, 184, 331n29; influence on Field, 228–229, 234–236, 344n16; planning for, 162, 163–164; racial organization of displays, 171, 172–173, 176, 178–179, 180, 186–187, 236; specimen collection for, 165–171, 179, 300n22, 331n27
Hygiene and demography, 117–118, 319n98
"Hygiene," eugenics and, 137, 319n98
Hyrtl Collection, 142, 148–149
Hyrtl, Joseph, 148

Indiana Medical History Museum, 320n1
Indigenous people, generally: at Louisiana Purchase Exposition, 136–137; pushback against collecting from sacred sites of, 255–257, 275; at St. Louis World's Fair, 63–65. *See also specific groups*
International Association of Medical Museums, 320–321n1
International Congress of Eugenics (ICE), 118, 119, 122
International Congress of Prehistoric Anthropology and Archaeology, 118
International exhibitions, generally, 134–137; new museum creation and, 159–160. *See also specific exhibitions*
International Symposium on Early Man (1937), 248–251

Ishi (Yahi Indian man), 94–101, 112, 316n64

Jacknis, Ira, 296–297n22
Japan, 31, 115–116, 244–245
Java Man, 246, 331n27
Jefferson, Thomas, 26, 301n29
Journal of Health and Physical Education, 211
Judeo-Christian narrative, ancient history and, 201–203

Keen, W. W., 153
Keith, Arthur, 57, 238–240, 249–250, 266
Kellogg, John H., 118
Kimball, James P., 31
King, Richard, 190–191
Kishu (Eskimo man), 97, 316n57
Kluckhohn, Clyde, 341n89
Kroeber, Alfred, 10, 105, 123, 185–186, 336n14; eugenics and, 223; Ishi and, 94–101, 316n64; as "primary collector," 109; as teacher, 315n52
Kroeber, Theodora, 315n53
Krogman, Wilton M., 339n55
Kuklick, Bruce, 296n22

Laboratory of Anatomy and Physical Anthropology, at Howard University, 188–191, 218–223
Laboratory of Anatomy and Physical Anthropology of Howard University, 1932–1936, The (Cobb), 188
Lake Okeechobee, Battle of, 91
Lamark, Jean-Baptiste, 57

Lamb, Daniel Smith, 218–219
Land of the Cliff Dweller, The (Chapin), 47
Lane, Franklin Knight, 90
"Last Great Congress of the Red Man," at Trans-Mississippi Exposition, 64
Laufer, Berthold, 234, 237–238
Leakey, Louis and Mary, 248
Lee, Ronald F., 299n9
Leikind, Morris, 302n35
Lindgren, Laura, 322n16
Little Big Horn, Battle of, 30, 302n38
Living exhibitions, 44, 63–65, 310n119
Looters and looting, 6, 21, 35, 38, 70, 74, 78–80, 84, 86–87, 90, 166–168, 200
Los Angeles Herald, 87
Louisiana Purchase Exposition (1904), 136–137
Lummis, Charles F., 47, 306n82

MacCurdy, George Grant, 118, 199, 337n33
Magdalenian Girl, 231–232, 241
Mascré, Louis, 175–176
Mason, Charlie, 40–41, 43
Mason, J. Alden, 271
Matthews, Washington, 83–84
Mayo, Charles, 333n60
McCown, Theodore, 249–251
Medical education: Army Medical Museum and, 29, 32, 302n35; early bone rooms and, 4–5; at Howard University, 188–191, 218–223;
museums and, 126–128, 130–131, 133–134, 144, 145, 154
Medical illustration, 133–134, 322n16
Medical museums, 14, 140, 320n1; competition and cooperation between, 146–147; contrasted to natural history museums, 145–146; legacy of, 154–157; lost limb collections from Civil War, 131–133; medical education and, 126–128, 130–131; 133–134, 144, 145, 154; medicine versus anthropology at, 149–151; public curiosity about and concerns about appropriate audiences, 127–128, 133–134; public health concerns and, 127, 130, 134–135, 137, 155–156, 323n28, 327n87; specimen preservation and, 126, 321n3; traits of, 145–146. *See also* Army Medical Museum; Mütter Museum
Menand, Louis, 297n24
Men of the Old Stone Age (Osborn), 334n70
Meringolo, Denise D., 313n27
Mesa Verde National Park, 40–41, 86, 90, 313n27
Metcalf, Ralph, 211
Mexican-American War, 91
Mexico, 193
Micka, Frank, 178, 235, 236, 332n48
Mitchell, S. Weir, 132–133, 150, 327n75
Mongolia, 103, 165, 169–170
Monogenesis, 24, 56–57, 208
Montague, M. F. Ashley, 275
Morgan, Lewis Henry, 22

Morton, Samuel George, 17–18, 23–26, 91, 148, 202, 300nn12,15
Mount Carmel fossils, 249–251
Mount Holyoke College, 260
Muller, Alfred, 2–3, 10, 293n2, 294n7
Mummies: conditions when found, 18–19; from Aleutian Islands, 18–19; from Egypt, 17, 37, 38, 40, 257, 262–263; prehistory and, 230–232, 261, 343n8. *See also* Cliff dwellers, of American Southwest
Musee de l'Homme, 199
Museum of the American Indian, 183–184, 255–256
Museum of Anthropology, at University of California, 95, 99–100, 101, 112, 225, 298n35, 307n84
"Museum period," 230, 268
Mütter Museum, Philadelphia, 9, 155, 280, 320n1, 321n9, 325n54; American Giant and, 140–141, 149; Chang and Eng Bunker and, 138–140; collections of, 128, 130, 142, 143–144, 150, 325n52, 327nn75,76; contemporary practices of, 282, 284; cooperation with Army Medical Museum, 146; establishment of, 129–130; growth of, 144–145; medical education and, 130–131; new facilities in 1900, 152–153; prehistory and, 149–151; purchase of Politzer auditory apparatus, 135, 148; racial classification and, 147–149, 151; specimen acquisition and racial pathology, 140–143, 325nn52,54; value questioned by College of Physicians, 143–144; visitors to, 144–145, 152–154, 156
Mütter, Thomas Dent, 129–130, 141

National Association for the Advancement of Colored People (NAACP), 189
National Academy of Sciences, 269
Nationalism, collecting and, 198–201, 202
National Museum of Health and Medicine (NMHM), 287, 320n1
National Museum of Natural History (NMNH), 257, 281
National Museum of the American Indian Act (1989), 278
National Park Service, 88, 90, 298n31
Native American Graves Protection and Repatriation Act (NAGPRA) (1990), 278–279, 283, 298n35
Native Americans: contemporary collecting and, 283; early collection of remains and research into racial origins, 16–21; Hrdlička's collecting and, 168, 170; living exhibits and, 64; military collecting and, 30, 31–32, 302n45; Morton's racial theories and, 24, 25; NAGPRA and, 279; Otis and, 5–6; prehistory and, 102–103; pushback against collecting from sacred sites, 255–257, 275; racial classification in *Story of Man* and, 177; remains still in museums, 15; repatriation of remains of,

277–280, 282; salvage anthropology and, 69–73; skulls at Mütter Museum, 150, 327nn75,76; treatment of remains on frontier, 1–3
Natural history museums, 12, 14, 17; contrasted to medical museums, 145–146; pathology focus of, 147, 326n68
Neanderthal remains, 121, 201, 240, 242, 246, 250
Negros. *See* African Americans
Nelson, Edward W., 253
Newman, Marshall T., 258
Newton, Philip, 165, 169, 329n17
New York Herald, 85
New York Times, 18, 19, 20–21, 138, 168, 267–268
North America, research into peopling of, 251–255
North American Review, 37

"Old Americans," 119–122, 177–178, 319n110, 331n40
Old Americans, The (Hrdlička), 119–120
Olympic Games (1932), 211–212
Omaha people, 111
Osborn, Henry Fairfield, 119, 334n70, 344n16
Otis, George A., 5–6, 33–34
Owens, Jesse, 212
Oxford University, 233, 237, 260

Paleoanthropology, 248, 275, 280, 337n33
Palestine, fossils from, 249–251

Panama-California Exposition, San Diego (1915), 91, 106–107, 158–159, 238–239. See also *Story of Man through the Ages*, at Panama-California Exposition
Peabody, Charles, 118, 201
Peabody Museum of Ethnography and Archaeology at Harvard, 244, 298n35
Pearl, Raymond, 188
Peking Man, 247, 346n48
Pepper, George H., 320n117
Peru, 50, 193, 205–206, 257, 263, 305n72; collection exchanges and, 122; hair samples from, 261; Hrdlička's collecting in, 106–108, 165–168, 179, 317nn78,81, 330n22; Stewart's collecting in, 107–108; trepanation of skulls from, 149–150, 331n27, 334n69
Phantom limb phenomenon, 132
Philippines, 31, 78, 165, 169, 329n17
Phrenology, 30–31, 146
Physical anthropology: changes in, 269–272, 274–275; cliff dwellers and, 17; cooperation within anthropology and, 350n102; displays at World's Columbian Exposition and, 44–51; eugenics and, 118–120; exposition displays and, 136; Hrdlička and, 56; previous scholarship, 12–13; racial classification and environment, 58–61; shift to universities from museums, 12, 209–210, 263–268; status in 1898, 53
Piltdown Man, 242

Plastination, displays and, 287–288
Politzer, Adam, 135, 148
Polygenesis, 24, 56–57, 208, 300n17
Positivist science, 7, 57–58, 160, 161
Powell, John Wesley, 22, 28
Prehistory and human origins, 227–276; American Association of Physical Anthropology and, 215–218; ancient history and biblical narratives, 201–203; Boas and, 60; collections and study of, 101–104; dating human occupation of North America and, 251–255; human evolution and, 246–251; at Field Museum exhibition, 232–243; indigenous people's pushback to collecting from sacred sites, 255–257, 275; Mütter Museum and, 149–151; physical anthropology's changes, 269–272, 274–275; physical anthropology and shift from museums to universities, 12, 209–210, 263–268; public's increasing interest in, 230–232; racial classification and, 243–245; racism and politics and, 244–245; shift from racial classification to, 8–9, 12, 74–75, 101–104, 192, 227–230, 272–276; Stewart and, 257–259; study of soft-tissue, blood and hair, 259–263, 349n82
Provenance concerns, 66, 80, 99–100, 142, 206, 302n33,38
Public health campaigns: international exhibitions and, 134–135, 137, 323n28; medical museums and, 127, 130, 134–135, 137, 155–156, 323n28, 327n87
Public interest, in exhibits: Army Medical Museum and, 27–28, 133, 301n30; cliff dwellers and, 16–21, 39–40; commercial exhibitions, 36–38; contemporary, 284–285; Egypt and, 17, 36, 230; Mütter Museum and, 144–145, 152–154, 156; prehistory and, 230–232; Smithsonian radio programs, 272–273; visitors thought to be repelled by but attracted to remains, 238–239, 286–287. *See also* Tourists, collecting by
Putnam, Frederic Ward, 21–22, 44, 111

"Race and Runners" (Cobb), 211
Racial classification: Army Medical Museum's call for Civil War military materials and, 28–34, 302n33; cliff dwellers and, 38–39, 41–42, 304n61; displays at World's Columbian Exposition and, 44–51; early collections and reinforcing of ideas of, 3–8; evolution and, 56–61; Field Museum's 1933 exhibits and, 228–230, 232–243; as focus of early research into cliff dwellers, 16–21; Hrdlička and, 54–56; medical museums and, 127–128, 146–147; Morgan's cranium studies, 23–26; Mütter Museum and, 147–149, 151; organization of collections and, 61–63, 75–83; shift to prehistory and human origins

from, 8–9, 12, 74–75, 101–104, 192, 227–230, 272–276. *See also* Scientific racism
Racial history, 103–104, 109, 230; ancestry and, 74; Dixon and, 203–209; Hrdlička and, 173, 183
Racial History of Man, The (Dixon), 204, 206
Racial mixing, Hrdlička on, 196–197
Rahall, Nick J., 278
Railroads, tourists and, 87, 90
Rameses II, Barnum's proposal to display, 36–37, 343n8
"Recent Criticisms of Physical Anthropology" (Boas), 58–59
Relativism, 161–162, 191–192
Repatriation of remains, 10–11, 277–280, 289; of Sioux warrior, 277–278, 295n12
"Repatriation: The Bitter End or a Fresh Beginning?" (Thomas), 279
Reversion, concept of, 121
Roberts, Frank H. H. Jr., 251, 254
Rogers, Spencer L., 330n22
Roosevelt, Franklin, 244–245
Roosevelt, Theodore, 86, 89, 181
Royal College of Surgeons, England, 320–321n1
Russia, 169–170, 245
Rutot, Aimé, 175–176, 235, 334n70
Rydell, Robert, 328n1

Salvage anthropology, 69–73, 123
Sand Creek Massacre, 30
San Diego Museum of Man, 9, 158, 165, 181, 183, 186, 228, 234–235,
330n22, 332n55; influence on Field, 228–229, 234–236
San Diego Union, 182–183, 185
San Diego World's Fairs and Southwestern Memory, 1880–1940, The (Bokovoy), 328n1
"Save Our Skulls" campaign, of Mütter Museum, 282
Scheper-Hughes, Nancy, 96
Schück, Adalbert, 171
Science News-Letter, 240
Science of Man exhibit. See *Story of Man through the Ages*, at Panama-California Exposition
Scientific racism, 188–226, 285; African Americans' physical constitution and, 212–215, 339nn55,63; American Association of Physical Anthropology and, 215–218; American Museum of Natural History and, 193–194; American School in France for Prehistoric Studies, 198–201; ancient history and, 201–203; anthropometric measurement and, 194–198; athletic ability and race, 211–212; Dixon and racial history, 203–209; eugenics and, 191, 192, 223–225, 341nn87,88; medical education at Howard University, 188–191, 218–223; physical anthropology at universities and, 209–210. *See also* Racial classification
"Scopes "Monkey Trial," 209, 228
Seminole War, 91
Setzler, F. M., 254

Sex-education, public health and, 323n28, 327n87
Shapiro, Harry Lionel, 224
Siberia, interest in, 102, 103, 115, 165–166, 168
Sioux Indians, killed in 1864, 1–3, 35, 293nn1,2, 294n9; Boas's measurement of, 46; repatriation of remains of, 277–278, 295n12
Skull collections: Army Medical Museum and, 27, 31–32, 35; of Morton, 22–26, 300nn12,15; Mütter Museum and, 130, 142, 149–151, 325n52; shapes related to intelligence and behavior, 195
Skull Wars: Kennewick Man, Archaeology, and the Battle for Native American Identity (Thomas), 316n64
Slave trade, Cobb's study of African American ancestry and, 221–223
Smithsonian Institution, 3, 10, 112, 151, 165; American Antiquities Act and, 86; anthropology department at, 51–53, 129; Army Medical Museum collection acquired by, 34–36, 51–53, 129, 146, 303nn53,54,55; dating human arrivals in North America and, 252; Densmore's Chippewa finds sent to, 70–72; displays in early twenty-first century, 280, 281–282; Dixon's criticism of Hrdlička's management of collections, 204–206; eugenics and, 117–119; Hewett's plans for Panama-California exhibition and, 162–164; Ishi and, 112; NAGPRA and, 279; organization and display of collections, 75–83; radio programs of, 272–273; reluctant to part with skeleton collection, 146–147; *Story of Man* specimens and, 165–166, 169–170, 177, 180; visitors to, 284–285; war opportunity and, 92. *See also* Hrdlička, Aleš
Some Strange Corners of Our Country (Lummis), 47, 306n82
South Africa, 170–171, 214, 217, 246–248
South America, 115, 116–167, 181, 193
Spencer, Frank, 13
Stewart, T. Dale, 102, 254, 262; Peruvian specimens and, 107, 261; physical anthropology and, 229–230, 257–259, 263–264, 266–267, 270–271
Stillé, Alfred, 130
St. Louis World's Fair (1904), 63–65, 164
Story of Man through the Ages, at Panama-California Exposition, 220; aging shown in, 176–178, 180–181, 186–187; busts, casts, masks, and artistic renderings used in, 159, 161, 165, 169–170, 172–180, 183, 186–187, 334n68, 344n16; eugenics and, 182–183; influences of, 186–187, 334nn68,69,70; planning for, 162–164; public and professional reactions to, 181–186, 333n60; racial classification, and displays by room, 171–181; racial classification, generally, 158–159,

160–162, 168–170, 177, 274; specimen collection for, 165–171
Strong, William Duncan, 271
Sullivan, Louis Robert, 46
Syphilis, 180

Taung Child, 246, 247
Thomas, David Hurst, 24, 279, 297n24, 316n64, 318n86, 348n71
"Thoroughbred" whites, 117, 177, 331–332n40
Thousand and One Afternoons in Chicago, A (Hecht), 210
Tolan, Eddie, 211
Tourists, collecting by, 27, 74, 85–87, 90, 133. *See also* Public interest, in exhibits
Trans-Mississippi Exposition (1898), 64
Trepanation, and displayed skulls, 149–150, 179–180, 322n51, 334n69
Trotter, Mildred, 260–261, 343n14
Tuberculosis, death from, 96, 97, 100, 137, 316n57
Tunis, Joseph, 325n52

Universities, shift of focus to from museums, 12, 74–75, 101–104, 209–210, 263–267
University Museum (University of Pennsylvania), 48, 50, 307nn83,84, 308n94
University of California, 11, 79, 94, 96, 113, 249, 251, 274; cliff dwellers and, 48, 307nn83,84; Museum of Anthropology at, 95, 99–100, 101, 112, 225, 298n35, 307n84
University of Chicago, 202, 209–210
U.S. Railroad Administration, 90

Valuing and exchanging, of collection specimens, 115–117
Vanishing races, 4–5, 8, 21, 72–73, 79, 84, 85, 110, 129
Virchow, Rudolph, 327n75
Vivisection debates, 322n14

Walcott, Charles D., 163, 168
Warren Anatomical Museum, Harvard University, 320n1
Washburn, Sherwood, 258, 267, 271, 274–275
Washington University, 214, 260, 344n14
Wetherill family, 42, 43, 48, 49, 86
Wetherill, Richard, 40–42, 43
Wilkinson, James W., 182–183
Wilson, Thomas, 208
Wissler, Clark, 193–194, 264–265
Worden, Gretchen, 327n76
World's Columbian Exposition (1893), 160, 227, 331n27; Barnum's proposal for, 36–37, 232, 342n8; cliff dweller display at, 47–50, 135–136; indigenous people as exhibitors and as research subjects, 44–47
World War I, 91–94, 195
Wormington, Hannah Marie, 347n62

CPSIA information can be obtained
at www.ICGtesting.com
Printed in the USA
JSHW020234090123
35813JS00003B/76/J